detail

EXCEPTIONAL JAPANESE PRODUCT DESIGN

ANDREW DAVEY

LAURENCE KING PUBLISHING

Published in 2003 by Laurence King Publishing Ltd
71 Great Russell Street
London WC1B 3BP
United Kingdom
Tel: +44 20 7430 8850
Fax: +44 20 7430 8880
e-mail: enquiries@laurenceking.co.uk
www.laurenceking.co.uk

A catalogue record for this book is available from the British Library

ISBN 1 85669 318 X

For Annie and Joey

Designed by TKO Design
Art direction: Andrew Davey
Graphic design: Rochelle Kleinberg

Printed in China

right Canon Powershot S40 digital camera
opposite NEC PaPeRo personal robot
overleaf Sony Aibo ERS-220

contents

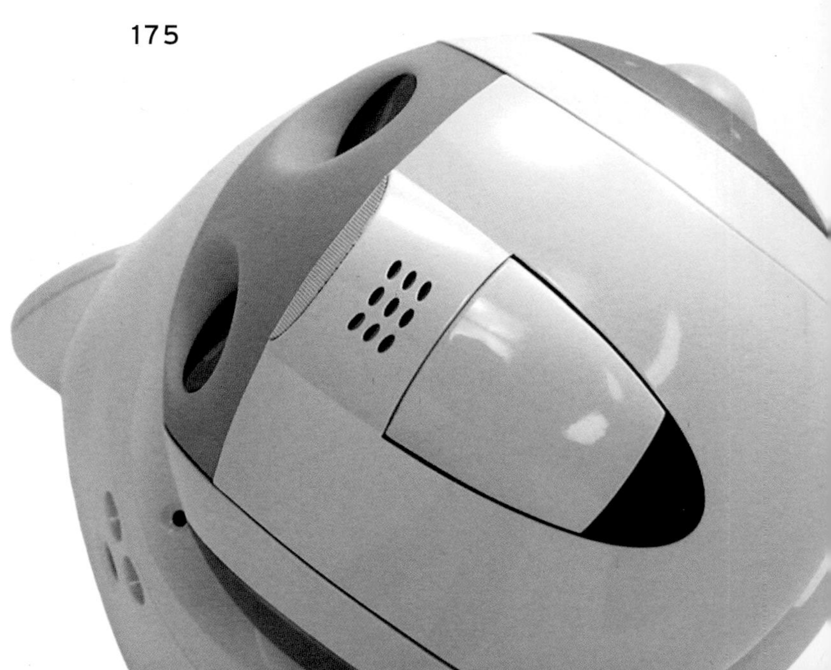

foreword

BY EIZI HAYASHI

To the ancient Japanese, China was known as Tang and India Tenjiku, and from here Japan eagerly learnt about religion, philosophy and law. Technologies, however, came through Korea. Art and science were studied and developed as Japan made constant efforts to improve. After the long seclusion from the outside world, the Meiji Restoration in 1868 sent out missions to observe the systems of politics, economics and education in Europe and America, and rapidly acquired the basis of a modern industrial nation. Meanwhile, in Europe the influential *Le Japon Artistique* was published between 1888 and 1891 by an exponent of art nouveau, Samuel Bing. He held the art of the Edo period in high regard. Japanese art was a revelation to artists and designers in the West, of which Van Gogh was one. Living in a country without natural resources, Japanese people have inherently developed an enterprising spirit and strong curiosity. It is this that has generated interest in the so-called Japanese spirit within Western learning. This intensified in the effort of developing the nation's wealth and military strength up to and during the Second World War, and the post-war reconstruction over 50 years after Japan's defeat, and now into the new century. East meets West, and vice versa, and so synchronicity and history recur.

Ten or so years ago, an English design director visited AXIS Inc to carry out market research. He was a Royal College of Art graduate who had become interested in Japan. So that was how, in my fourth floor office, I first met the author, Andy Davey. He told me he wanted to see many examples of Japanese product design, and to visit and seek cooperation with manufacturers unknown in the UK. He left me with a gift of a sectional cardboard snow-scraper, and on the following day the first snow fell in Tokyo. Every year flowers blossom anew, but we never remain the same. We all grow and change year by year and decade by decade. Every encounter is significant, for it will never recur. It happens in a flash and affects one's life in many ways, good or bad, and every one of us is responsible for the happiness that comes with it. Without discernment, it is not easy to make sound judgements and without cultivating virtue, others will not come with you.

Andy Davey is hard working, sincere and genuine. He has been working with Japanese companies for years, gaining a deep view of their nature and widening his circle of friendships. Here, he has put together his knowledge and experience. It is said in Japan that sitting on the same rock for three years will bring enlightenment, and at the same time the stone becomes warmer! How many times has Andy visited us in Japan? With this tremendous passion his personal and working relationships with Japanese people have been expanding, and his understanding of the industry and its merchandise has also become deeper. He has already managed to climb to the top of Mount Fuji.

It is a hard and long road to achieve success in product design. By going further and deeper it is clear who has actually promoted R&D, who has made the right decisions at the right times and who has brought the project into fruition in the face of 98 per cent opposition. I hope the truth will come out! Publishing the real story is truly a literary gem.
Go for it, Andy!

26.07.2002, Suzushin, Arakicho, Tokyo

OLYMPUS
ON
OFF

introduction

Throughout history, man has defined himself by the tools he has created and the progress they have generated. Social advancement is still measured in these terms today. As consumers, our development is as much influenced by the products we use as by the events we experience. My generation, born in the 1960s, was perhaps the first to embrace fully the optimistic future that technology promised, as moon shots, supersonic flight and the semiconductor became icons of endeavour in an increasingly product-orientated society.

Technology has a firm grip on all our lives. The source of this technology in the form of everyday 'tools' for life is predominantly Japan. For better or worse, our man-made world is becoming more and more conditioned by fewer and fewer mega-organisations, and many of these are Japanese. Personally, I recognise and value the crucial influence that Japanese design and products have had on my life, not only on my product-cluttered youth but also on my career as a designer.

left limited edition Olympus Écru 35mm camera

I recall clearly the beginning of my fascination with Japan, both country and culture. I was at primary school, at the tender age of ten, when my class was assigned a project to study individual countries. I was given Japan, and of all the embassies contacted to request information, it was the Japanese one that sent the most enviable quantity and quality of material. Huge posters featuring images of the Shinkansen (Bullet Train), as a blue-and-white blur shooting past a snow-capped Mount Fuji framed with cherry blossom, are every bit as evocative now as they were then, and a powerful image of the juxtaposition of modern and traditional that has always been identifiable in Japanese culture.

Throughout the 1970s my bedroom was filled with Japanese toys and gadgets. These products reflected my perception that things from Japan were modern, flash, affordable and compact all at once. In my early teens, it was coolest of cool to own a Yamaha FS1E, a 50cc moped with motorcycle pretensions. This model single-handedly captured the imagination of a generation of future motorcyclists; my elder brother bought one, and grew in my estimation by his choice. Many of the older school leavers bolstered their credibility by jumping from moped to the next road tool of choice, the Honda 400/4, a four-cylinder dream machine that was, inevitably, a mere stepping stone to the brilliant Honda CB750.

While riding (and falling off) my brother's borrowed moped, I had to take care not to scratch my LED watch, a huge ingot of stainless steel and deep-red glass. To tell the time, I had to press and hold down a button on the side of the watch and squint at a tiny readout of indecipherable digits - not an easy task on the move. I recall a friend arriving at school, his sleeve permanently pushed up to reveal one of the first, and very covetable, Seiko LCD watches, a magical technology then. Alongside the watch, many other Japanese products were admired and collected: a '76 Suzuki TS250 motorcycle, a blue metal TPS-L2 cassette player (soon to be known as the Sony Walkman), and a Pentax MX camera, perhaps my first truly creative tool.

The direction and output of a designer is influenced by personal and collective experience, while growing and developing. This psychological luggage is most obviously collected from cultural sources - literature, film and television - but also through the physical and visual experience of artifacts and products. In the 1970s and 80s, my generation was the first to experience the impact of the wave of new products from Japan.
Most were painfully desirable products, innovative and unique to their age, such as the Walkman, Zoids and Game Boy. In the 1990s, Tamagotchi and PlayStation became icons on the global product landscape. Each of us has memories of products that had a certain significance in our lives, such as our first digital watch, calculator or video game. These products have helped shape our attitudes to technology and its place in the modern world.

So much has been said and written about Japan and the Japanese, yet relatively little is really understood by Westerners. Beyond the stereotype and cliché, Japan remains a country wreathed in misinterpretation and myth, exoticism and exaggeration, which is unique in a major industrialised society. Despite existing in an age of global information and international consumerism, Japanese culture remains oddly oblique and occasionally

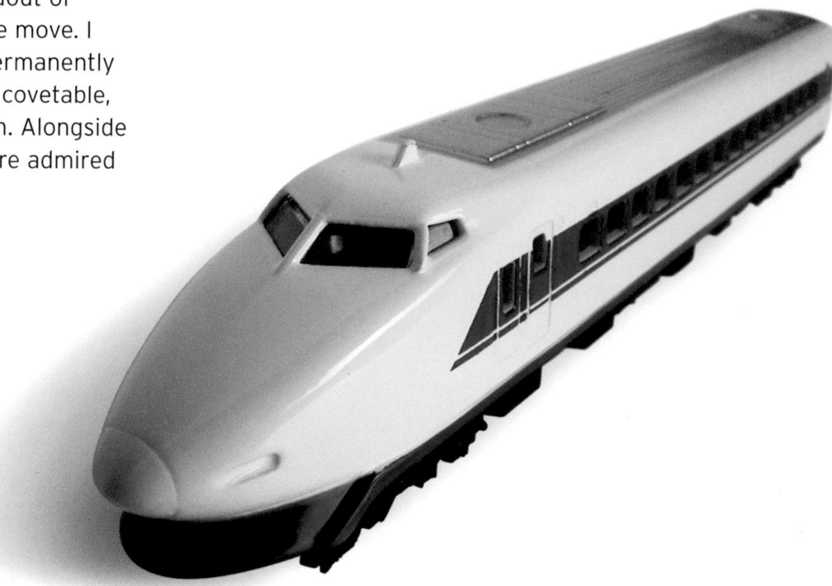

right Shinkansen Series 100 model

unfathomable, both challenging and captivating, when observed from outside. Even those more familiar with the nation's culture find it complex and, at times, contradictory, yet always fascinating. Some maintain that the complexities and contradictions may be overplayed, perhaps consciously, to create a useful cloak of inscrutability in international business and politics. What is certain, however, is that more information about Japan is available to us now in all forms of media, resulting in a greater interest in this complex culture.

This book aims to throw a spotlight on a narrow but economically crucial part of Japan's cultural life – its design and manufacturing virtuosity – that graphically illustrates the modern period of regeneration in a creative manufacturing society that has no direct equal. We are, of course, familiar with many traditional representations of Japan that neatly embody those oft-quoted qualities of hard work, persistence, teamwork, rigour, manners and beauty, especially in the visual sphere: calligraphy, ceramics, the tea ceremony, flower arranging, the martial and performing arts. Today, however, the Western media increasingly focus on the conflicts and contradictions that result from a culture that is still, to a certain extent, feudal and patriarchal. Where once the individual was subjugated in favour of the collective, the recent erosion of the post-war concepts of 'a job for life', financial security, social responsibility and teamwork has meant that the creativity and identity of the individual is becoming an issue of growing importance in business and at home.

above Sony TPS-L2 cassette player
right Pentax MX camera

One of the most written about and discussed aspects of Japan's 20th-century history is its rise to economic pre-eminence since the Second World War. Put simply, it is a unique case study of the power of national teamwork, perseverance and commitment to research and development. Then, after 40 years of continuous economic growth, the ultimately unsustainable economic 'bubble' burst in the early 1990s, and the resulting slow-down and recession have forced many in Japan and the Asian region to question their social and economic orthodoxy and consider significant structural reform. In some areas, in particular the political, the acceptance of change has proved painfully difficult and slow. Japan may have acknowledged the need to look at and learn again from other economic models and to renew its structures and systems; but instigating change in a culture still heavy with political denial is neither easy nor straightforward. There is a general awareness of the need to shed the instinct to conform and the *mura ishiki* or 'village mentality' on which traditional society rested, and to adopt a more individual approach to empowerment and enterprise. But at what cost?

While Japan's ability to adapt to changing circumstances is what has made it such a fascinating culture in the post-war period, the past decade of post-bubble economy has led many Japanese to question the country's direction and future place at the global economic table. Japan is still comfortably the second biggest economy after the US, but many people are beginning to count what has been lost from their culture along the way to economic superstardom as well as what has been gained. Within Japan itself, some see the shift from the ideal of teamwork and the 'greater good', to a form of self-assertion and individualism as a negative thing. Others, particularly younger Japanese, find the gradual loosening of the strictures of traditional society a positive and invigorating change for the better.

Technology

The application of technology in our physical and emotional lives has produced undeniable benefits in terms of entertainment, health and contentment. Technology and its uses, for the most part, are a modern-day elixir, a constant and controllable source of wealth and of

this page from top Nintendo Game Boy, Bandai Tamagotchi, Sony PlayStation 2

wonderment, which sustains our market-driven economies and cultures; a rich and satisfying demonstration of man's ability to confront and overcome problems and to supply solutions, even to previously unidentified dissatisfactions.

Yet it is also argued that technology in the modern world has negative outcomes elsewhere on the globe, most obviously in terms of the disparity in wealth and knowledge. US, European and Japanese companies are responsible for registering over 90 per cent of the world's patents, and while the US leads the world in the development and delivery of military technology, Japan's pre-eminence is in consumer markets. Consequently, designers and engineers in Japanese corporations have highly refined tools and the expertise frequently not found or available elsewhere, to develop advanced new products in many market sectors, and a knowledge and innate understanding of how to deliver innovative technology to a hungry global consumer.

Consumers and Consumption

Japan has always enjoyed a loyal as well as enthusiastic domestic consumer base, and one with all the wealth and dedication to shopping culture that a nation (with hardly any natural resources to speak of) could possibly want. In fact, one of the contributory factors in Japan's emergence as an economic powerhouse during the past 50 years has been the exhortation from its post-war governments to buy Japanese goods for the benefit of Japan, a notion dutifully accepted by the consumer.

Despite the outward appearances of a typical Japanese shopping experience - subservience to the customer, personal attention and serious gift-wrapping - the power of consumerism lies firmly in the hands of the mega-corporations who control distribution, availability, prices and, of course, choice. Or rather, they did until recently.

As a result of the fragmentation of Japanese work culture, caused by the nation's economic crisis, the Japanese consumer now wears the expression of a wily and knowing shopper, no longer wide-eyed and enchanted by the glittering prizes of the 1980s, but looking carefully and critically for innovation, quality and life-enhancing solutions. Japanese consumers, among the most sophisticated and technologically astute, have become the most unpredictable and discriminating. This cultural and economic shift has had a dramatic effect on manufacturing companies and their perceptions of the vast domestic market of 126 million Japanese citizens. Some corporations have resorted to introspection, being concerned with local markets, while others realise that restructuring and the continued production of world-class products remains the only way forward.

Designers

Traditionally, it has been usual for Japanese manufacturing organisations to use a rigid system for the development of a given technology, based on the methodical, rational application of intelligence to a specific issue or problem. This structured approach encouraged

right NEC PaPeRo personal robot

corporations to evolve new products and product innovations rather than take sudden, revolutionary leaps. Until relatively recently, most Japanese designers worked like this, in-house, as part of an anonymous, corporate team, without individual recognition. However, as a celebration of excellence, it is an important function of this book to acknowledge the work of Japanese designers and full project credits are included where possible. This is in line with the growing interest in the contribution of individual Japanese designers, both at home and abroad; interviews with and credits for the work of these individuals now regularly appear in international design magazines.

Information Technology

As in all developed countries, the irreversible tide of information technology is evident in Japan, where a wave of changes to the industrial and social fabric of the country has generally been considered as benign. As elsewhere, the internet in Japan is grasped enthusiastically as an enlightening and enabling tool for disseminating information, not just in conventional office or business situations but also in places like convenience stores (which are much more abundant in big cities like Tokyo), where internet terminals are provided to create a multi-function platform for banking, entertainment, gaming and e-mailing as well as shopping. Digital television has become a uniting force for communication and broadcasting systems, leading to dynamic links with mobile communications. Multimedia television services such as SoNet from Sony, accessible with multimedia handsets (2.5G and 3G or i-mode and FOMA), are capable of downloading and sending compressed video as well as e-mailing and texting, a growth area in which Japan leads the world.

The IT revolution has also provided opportunities for Japanese corporations to manufacture more effectively off-shore in other Asian, European, North and South American countries; sometimes for cost benefits, sometimes for more precise targeting within international markets. However, this globalisation of the product has raised questions in some sections of Japanese industry about what actually defines a 'Japanese product' today and how to communicate a pride in its production to the Japanese and global consumer. The acknowledgement of the role of the designer is one important way.

Environment

Products and technologies can become outmoded as quickly as they once appeared. Rather like a mayfly, the lifespan of most products is disturbingly brief and a good number of the current products featured in this book will, by the time it is published, already be redundant or replaced by newer, brighter versions, with even more potential to perform and please. Little can be bought without the disposal of something else; this is the reality of modern consumption. The new and shiny arrives while the old is rejected both physically and psychologically. Years of ignoring the environmental problem of the disposal of goods has led to an acute situation in Japan as elsewhere. Re-use may be a way of delaying the burial or cremation of a 'dead' product and there is a thriving industry in Japan of exporting used cars and of the repair, re-sale and export of used telephones, video players and televisions to developing nations. The main difficulty associated with this type of re-use economy is that it really only defers the ultimate disposal of the product, and to a country that may be without regulated disposal systems.

Re-use in Japan can also be seen in the emergence of concepts like 'Book Off', a well-stocked, comfortable environment that happens to be a second-hand bookshop, and its cousin shop called 'Hard Off', which sells used household goods. There are similar outlets for used golf clubs, restaurant equipment and audio-visual products, as well as more *ad hoc* sites like flea markets and antique shops. In these well-organised and socially acceptable second-hand outlets one can witness a change in Japanese consumers' attitudes to re-use, and a growing interest in the collectible. There are government initiatives designed to make the Japanese consumer as aware of the importance of recycling and of re-use as people in other regions, particularly Europe, but the framework for a social sea-change like this still requires the establishment of a recycling industry, as well as the building of greater social responsibility through financial incentives.

opposite Tomy 'Dark Spiner' Zoid before and after assembly

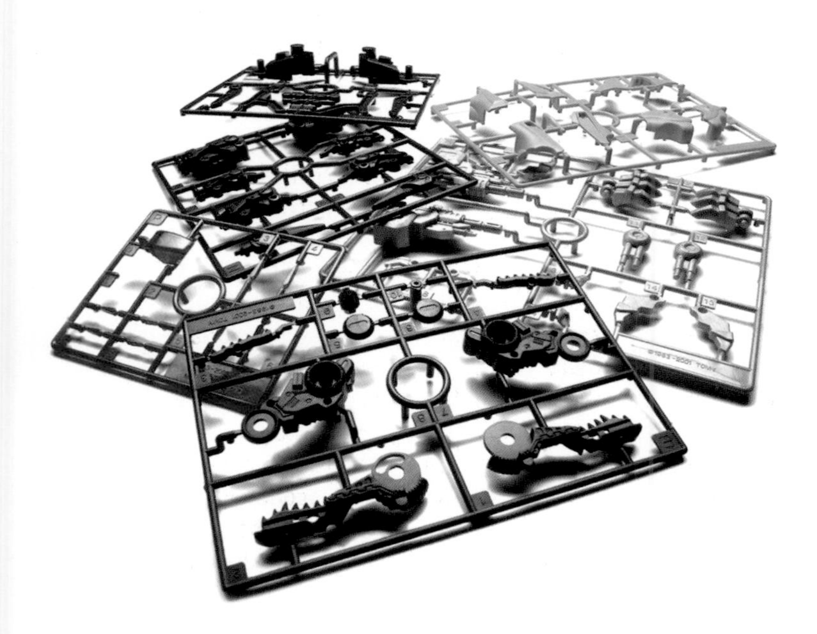

Discount stores in Japan are on the increase, and the rise of giant ¥100 (85 cents or 50 pence) shops are producing new and insidious environmental problems for the future. They stock almost every conceivable product that can be bought for ¥100, such as watches, gadgets, toys and food. Accompanying this phenomenon, however, is the concern that consumers will dispose of these goods in a more cavalier and thoughtless way than before. So cheap imported products (from countries where labour is cheap, but conditions poor) will add to the ecological burden of future generations.

Conclusion

By removing mass-produced, widely available products from their context as functional tools, we can better analyse the power of technology, of form and of the desire for the object. But we should not forget that most of the products featured here are manufactured in large numbers, to a critical cost, for a specific function. Yet each one remains a *tour de force* of design that will enhance, excite or enlighten our functional environment to a certain extent.

Detail elucidates many of the qualities found in traditional Japanese aesthetics – minimalism, elegance, simplicity, symmetry, workmanship and clearly-expressed function – to demonstrate the evolution of these qualities into modern, highly technical, mass-produced artifacts. Tracing developmental contexts through product case studies, it outlines how historical circumstances have informed current design trends in Japan, and how global observers and consumers have experienced a growing recognition of, and respect for, the creativity, originality and uniqueness of the Japanese design culture that permeates our modern world. It throws a spotlight on a selection of mass-produced Japanese products designed for our homes, offices and roads to reveal their precision and technical brilliance – products we interact with every day, but whose complexity of invention we only partly understand. Their technical perfection is sometimes so complete as to be almost invisible. This book aims to uncover and acknowledge their visual and technical power. In these instances, God truly is in the details.

1 artificial emotion

HEROES OF AN IDEALISED WORLD

This chapter explores Japan's obsession with a stylised world populated by manufactured heroes and fantasy characters; on television, in manga (comics) and in cinema – media that have grown enormously in popularity since the Second World War. From television characters such as Ultraman and Astro Boy in the 1960s, through to Zoids and the current explosion of robotic products like Sony's Aibo and NEC's PaPeRo, there has been a tradition of using technology to enhance our lives emotionally and functionally. More serious contemporary examples of an idealised modern world can be seen in projects like Honda's ASIMO, initiated by dreams of assisting, or even replacing, humans in menial or dangerous work. Sony's Aibo project taps into the more emotional side of the human imagination, demonstrating our capacity to be entranced and entertained by technology. Many other examples continue this Japanese tradition of man-made, culturally unique characters, including Bandai's Gundam series and Takara's Aquaroids.

One of the most vibrant sub-cultures to appear in the aftermath of the Second World War in Japan was the dynamic and inventive spirit that created a cast of new, super-realistic characters. Japanese film-makers, writers, artists and designers sought to dispel the chaos and desperation of everyday life in post-war Japan by replacing it with a world of imagination and fantasy. These early characters, some based on imported American models, others drawn from traditional Japanese storytelling, became icons of a new and brave era of progress and regeneration. Historically, Japanese culture has always produced real and imaginary heroes, such as the Seven Samurai, but these new and sometimes crude commercial heroes were completely of their time, in both action and imagination.

left Ultraman characters

The father of manga was Osamu Tezuka, who in 1947 created the distinctive and uniquely Japanese fantasy industry that thrives today – one-third of all publications in Japan are manga. Astro Boy first appeared on Japanese television in 1963 (the year before the Shinkansen was introduced and the Tokyo Olympics were held) and ironically, was syndicated for American television, such was his popular appeal. Astro Boy's imaginary 'birthday' is 7 April 2003, and this will be the occasion for commemorative celebrations all over Japan.

In 1966 the most popular of all Japanese fantasy characters arrived – Ultraman. Arguably a Japanese version of Superman, Ultraman changes from human to superhero and protector of mankind whenever a threatening opponent emerges. He transforms himself into a 40-metre (130-feet) tall semi-aquatic creature in a spangly silver suit, yet can only stay in powerful alien form for three minutes. For some, he has been the most influential fantasy character for an entire generation of Japanese designers and engineers, and while the notion of imaginary characters directly influencing and inspiring designers is claimed by some, others see it as an irritatingly reductive take on recent cultural history.

For a more contemporary influence on current designers, look no further than Gundam. At first glance, Bandai's Gundam products look like self-assembly playthings, plastic models of super-endowed, other-worldly aggressors just like any other styrene-kit fantasy figures, such as Zoids, Transformers and so on.

However, Gundam is more than that. Like many other cults in Japan, this set of super-charged characters has a deep and committed fan base consisting of all ages, who are drawn into this construction-kit world of tribal allegiances and hyper-realism. The Gundam universe spans television, film, websites, books and magazines. Rare and even hand-painted versions command premium prices, and the creators and artists behind the characterisations are fêted as demi-heroes themselves. Gundam is a cultural phenomenon right from the heart of Japan's tradition of mythical heroes and superhuman automatons. As influential in contemporary design as Astroboy or Ultraman before it, this culturally significant fantasy world is as much fun as it is beautifully detailed. While the influence of historical Japanese figures such as the Samurai weighs heavily on the conception and styling of Gundam, so does the culture of super-weaponry which is so popular among the youth of Japan (ironic in such a relatively safe, gun-free society). 'It's just for fun' seems to be the message here.

Using highly advanced plastics and finishes, high-tech moulding and assembly techniques, and of course, breathtaking detail, Gundam represents something singularly Japanese in both conception and realisation. Anyone familiar with styrene assembly kits from their youth will appreciate the sheer visual virtuosity and technical brilliance of the Gundam genre.

Ultraman's storylines were more complex than mere tales of saving the planet. Early Ultraman adventures were thinly veiled analogies of Japan's contemporary ills, such as the dangers of economic expansion, environmental pollution and traffic accidents. It could be argued that these characters, and Ultraman especially, taught a generation of Japanese children about the values of social responsibility, perseverance and teamwork, contributing to the strength and vivacity of Japanese industry in the post-war decades. This meeting of artifice and emotion can be found in another contemporary field of creativity – robotics.

Contemporary robotics in Japan are more advanced than anywhere else in the world, and robots are usually developed either for entertainment or for human assistance. The use of robots for mundane and tedious work began in Japanese car factories, and 75 per cent of robots now used in industry are made in Japan. Robots have always occupied an important position in the Japanese imagination, where the triumph of technology over evil is dramatised in so many manga tales and in film and theatre. Today, advanced, research-based robots can be found in many areas of Japanese life – on the street (vending machines), in transportation (Shinkansen), as entertainment (Aibo) and as human assistants (ASIMO).

There is also a strand of robotic endeavour that is located squarely in the domestic environment – build-it-yourself robots. A trip to any electronics market or specialised shop in the Akihabara district of Tokyo will provide you with all you need to assemble your own droid, robot or automaton for whatever purpose (or Mary Shelley fantasy) you may have. This cottage industry is worth ¥5 billion (US$42/£27.5 million), and has spawned an entire sub-culture of magazines, websites and meetings. There is even a grand robot exhibition held each year, called Robofesta. Here, industry, home roboteers and children happily co-exist in a fantasy world populated by robots, controlled by humans. It is also, however, a serious annual symposium where enthusiasts and company executives can rub shoulders and share ideas about the future of human/robot cohabitation. One of the most eagerly discussed applications is the domestic 'home-help' robot that could manage the home and care for its owner.

An ageing population is an increasingly important social and political issue in all developed countries, where birth rates are falling. This brings with it a range of problems, including a growing reliance on state welfare and health care, higher insurance premiums and a fall in tax revenues. Nowhere are these problems more acute than in Japan, whose population is ageing faster than anywhere else in the world. Apart from delaying retirement (Japanese people can expect to live longer than most, currently 83 years for women and 78 for men) and encouraging immigration (Japan still has one of the most ethnically homogeneous populations in the world), Japanese industry is trying to do its part to encourage a more youthful society. For example, car-maker Daihatsu offers employees' families ¥160,000 (US$1,300/£800) and free car rental for three years on the arrival of a fourth child, and toy-maker Bandai awards ¥100,000 to families producing a third. The Japanese government is considering more radical solutions, including initiatives to facilitate the 'import of youth and the export of age' by setting immigration targets of up to 500,000 a year and the retirement age above 65, both designed to make up the predicted shortfall in the work force.

Some other solutions to the problems of an ageing society lie in technology, specifically in robotics. Previously only the stuff of science fiction, the Japanese government plans to support actively the development of robots as a key industry for the 21st century and has earmarked ¥3 trillion (US$25/£16.5 billion) by 2010 to boost the domestic robotics market, rising to ¥8 trillion by 2025. The aim is to create robots that can effectively replace humans in dangerous jobs such as land-mine clearance but which can also assist in the domestic realm. Other familiar robotics scenarios include driverless trains, intelligent vending machines and cash dispensers. Looking further into the future, we may see robots performing medical operations and the use of invasive micro-robots. Olympus Optical is researching into diagnostic micro-robots that enter the body to search for and even to treat abnormalities. Redolent of films like *Fantastic Voyage* and *Inner Space*, the capsule endoscope is a diagnostic robot that has a diameter of 1.5mm and an integral lens of 0.5mm, and that can travel around the body via the bloodstream. Yet we are still in the early stages of the robot's development, and the revolution that the sector is waiting for is true artificial intelligence (AI), when robots will be able to think for themselves. This will be as

right Atomic Boy

important as the arrival of the transistor, the microchip or the understanding of the human genome.

There has been a significant shift in the attitude of the Japanese public towards robots during the past few years, and the appearance of domestic robots is now imminent both in terms of technology and psychological acceptance. A domestic world where a robot is as familiar as the postman has been brought closer by research into AI by some well-known companies, including Sony, NEC and Honda; companies for whom a huge R&D budget, a familiarity with domestic environments and sophisticated marketing tools are givens.

Currently there are many projects exploring the very limits of today's technology to create artificially intelligent robots for the mass market, to assist and entertain us. Robot design is taking centre stage in many large Japanese corporations and some observers believe that robotics may offer the right challenge and opportunity for Japan to re-establish its global industrial leadership, particularly if, at the heart of every 'smart home' of the future there will be a smart robot, a kind of technological Jeeves, functioning as the network appliance centre for the control of, and communication with, other electrical appliances.

The current iterations of the robot as a communication or interaction device, to assist, entertain or educate us, still require certain triggers for humans to identify or interact with them emotionally. Most domestic robots have an identifiable face with two eyes and a way to nod or acknowledge information. While technologically speaking this response behaviour is unnecessary, all robotics projects seem to follow this route; one which I believe is a dead end. A television does not replicate the human face, yet it is a two-way, interactive communication tool with its own variable but identifiable character that has been developed over many years. Robots need to develop a singular and personal character for themselves.

In today's optimistic and post-*Westworld* environment, there are two distinct strands of development in synthetic intelligence. One is where precision-engineered robots replace or assist humans in a purposely benign and unthreatening way (Honda's ASIMO). The other is the robot as commercial plaything, a companion and source of entertainment (Sony's Aibo). These two strands are important, as both Honda and Sony took care not to produce a robot which might threaten the user by awakening fears and insecurities about their worth or relevance in the real world (particularly in post-recession Japan). For instance, Honda's ASIMO is a pint-sized (1.2-metre/4-feet-high) human figure, the product of a highly serious research process, but presented rather like an ultra-bright kid from next door who is keen to do the housework. Sony's Aibo is a terrier-sized bundle of canine interactivity, far more intelligent than the average household hound, and only slightly less lovable.

For a more traditional kind of robot, the super-human-machine-as-aggressor so familiar from television cartoons, film and literature, one simply has to look at the toy market, where the clock has stood still since the 1960s – a time when you knew exactly where you were with a robot and that was as far away as possible. Companies such as Bandai and Takara produce ranges of fantastic, advanced and stimulating robotic characters, but ones which are firmly rooted in fantasy tales and their obsession with conflict and gladiatorial machismo.

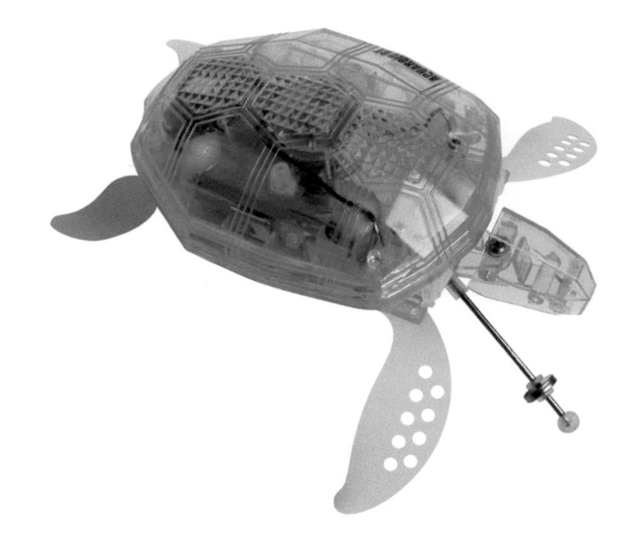

right Takara Aquaroid BT 03
opposite Bandai Gundam Neue Ziel (top), (bottom, left to right) EVA-01, EVA-02 and Hyaku-Shiki characters

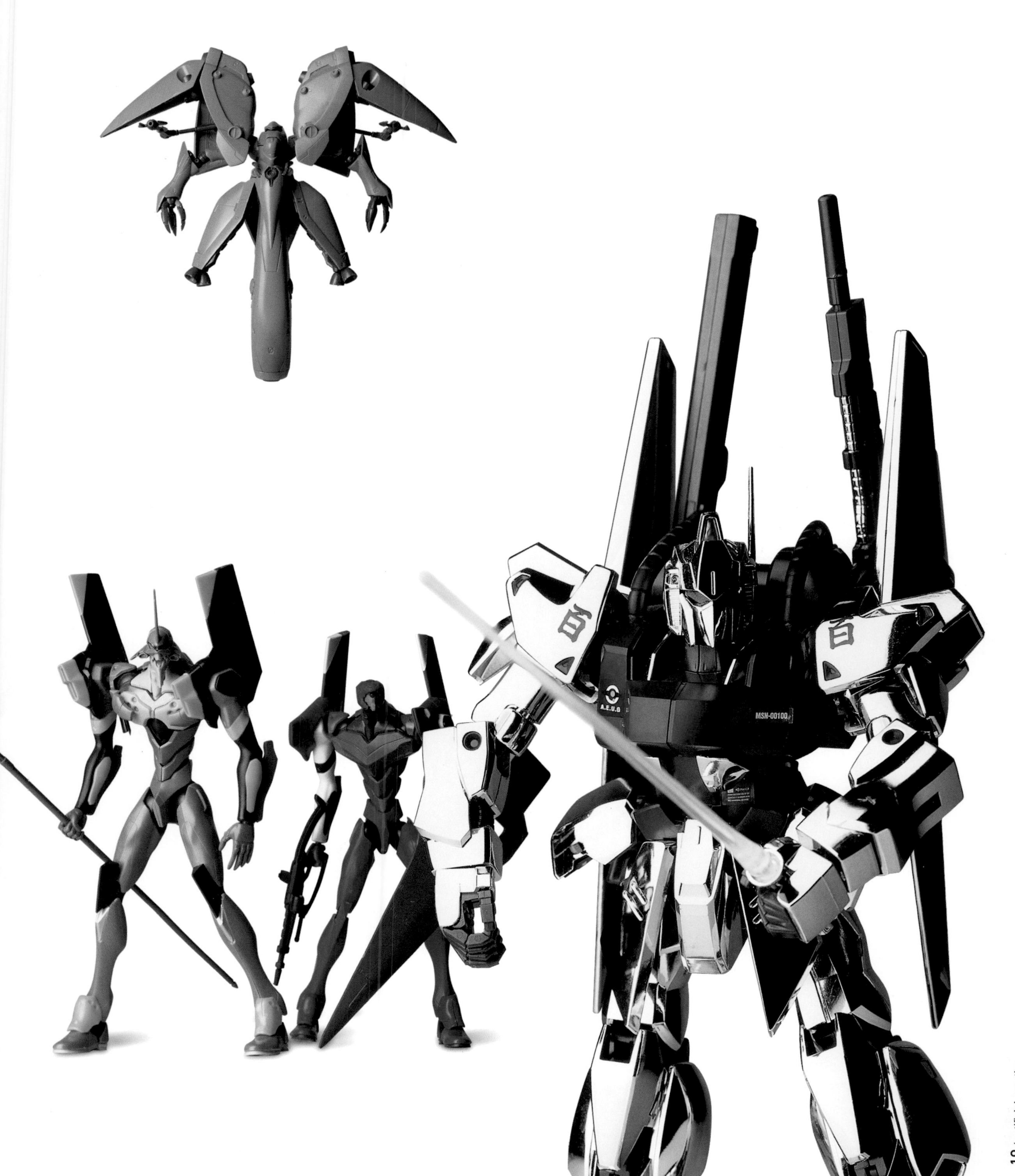
百
百
A.E.U.G
MSN-00100

Honda R&D Co., Ltd

ASIMO (Advanced Step in Innovative MObility)

Almost all major Japanese corporations invest heavily in R&D for their own product sectors. Many, however, invest in innovation for its own sake. Some of the most interesting research into robotic movement control has recently been undertaken by the Honda Motor Company.

Within the depths of the Honda Wako R&D Centre in Saitama, amongst the thriving engineering departments, the design studios and the research and planning sections, is a relatively compact but hugely influential part of Honda's future. Wako is the home of ASIMO, one of the most recognisable and certainly the most advanced humanoid robot.

ASIMO is the latest humanoid robot to be developed by Honda building on their previous research into autonomous locomotion with the P2 and P3 robots. It is the result of over 16 years of painstaking research and development and was conceived to function in real human environments, using intelligent, real-time, flexible walking technology. The robot can change direction in a fluid manner by predicting its next movement and shifting its centre of gravity in anticipation – the greatest advance so far towards replicating human motion patterns. Honda believes that this technology will allow robots of the future to work in closer harmony with their human masters, while negotiating tasks and obstructions on their own.

While Honda, like most global corporations, talks about company principles for the future such as 'glocalisation' and 'eco-solutions', it also speaks with conviction about people's dreams and a commitment to making those dreams a reality. 'The Power of Dreams' is Honda's mantra. This may sound like the usual corporate bluster, but Honda is producing work that really does suggest a realisation of the desire for a world populated by benign, intelligent machines. Company CEO and President, Hiroyuki Yoshino, famously said of ASIMO, 'One never tires of looking at a dream'.

opposite Honda's ASIMO

ASIMO
HONDA

The project that spawned ASIMO began in 1986, with the objective of creating a 'partner for the people', a robot that could have a function within everyday society. Initially, this ambition was limited to a two-legged humanoid that could walk, but the project developed into refining the mobility that would play a central role in a future where robots could assist mankind. For this goal to be realised, Honda needed to produce a two-legged technology that would allow robots to walk, climb stairs, predict and avoid obstacles and to walk on uneven surfaces. It was certainly an ambitious project in 1986.

Walking may seem simple to us, but to simulate walking successfully is complex in the extreme. The first ever walking prototype Honda produced, the E0 (E for experimental), could walk forwards, but only very slowly, taking up to five seconds between steps. Slow walking is relatively easy to achieve as the centre of gravity remains beneath the soles of the feet. Faster walking causes the centre of gravity to move according to the effects of inertia generated by the body. It became clear that faster walking was required if E0 was to cope with uneven surfaces, crucial for realistic mobility.

Between 1987 and 1991, research into the human ability to walk was carried out, together with studies of how primates and other animals walk. Of particular interest was the way in which joints are articulated and controlled by the body, and the role that the hips, knees, ankles, heels and toes play in the constant linear and lateral adjustments the body makes while walking. These observations were integrated into the robot's mechanical specifications. Humans also use the inner ear for balance, power from the muscles, and sensations from the skin to achieve the complex, but intuitive, nuances of movement. To replicate these subtle, yet crucial details, gyroscopes, electronic angle and force sensors, and 6-axis joints were required. Man also has the ability to absorb shocks and impact: for instance at walking speeds up to 2-4kph (1-2.5mph) the impact is 1.2-1.4 times the body weight, and at fast walking speeds of 8kph (5mph) the load increases to 1.8 times the body weight. To counter this problem, impact-absorbing materials and compliance controls were incorporated into the specification. The E2 prototype achieved a speed of 1.2kph (less than 1mph) on a flat surface. It had a strangely unnerving appearance, a truncated torso with a scary gait complete with visible cables and metal components, in fact a very *Westworld* incarnation, a film where the benign robots developed for the amusement of humans eventually over-throw their masters.

Between 1991 and 1993 various prototypes followed the E2, each one possessing greater walking speed and stability. The P1 robot (P for prototype) could carry out certain functions like turning light switches on and off, opening

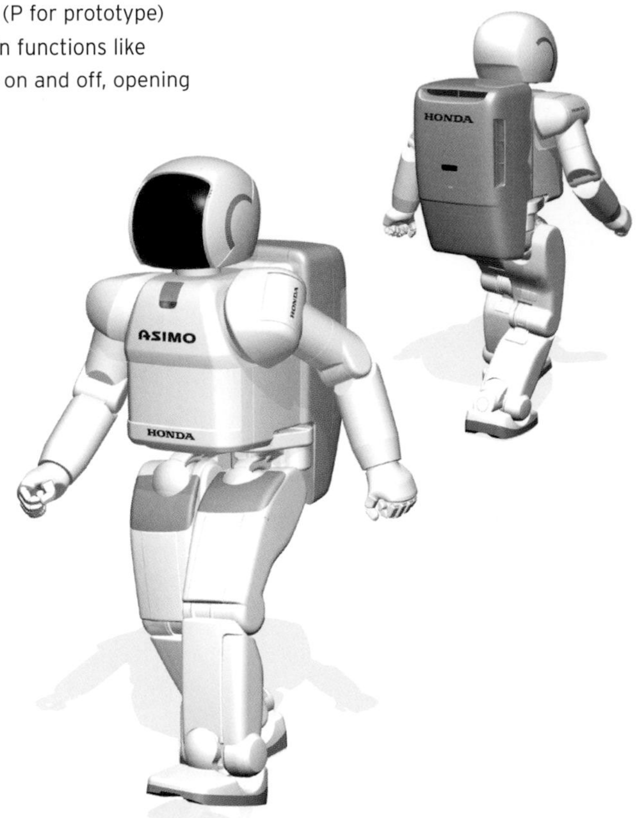

right ASIMO CAD views
opposite top evolution of Honda's robot programme
opposite bottom (left to right) ASIMO, P3, P1, E5, E0

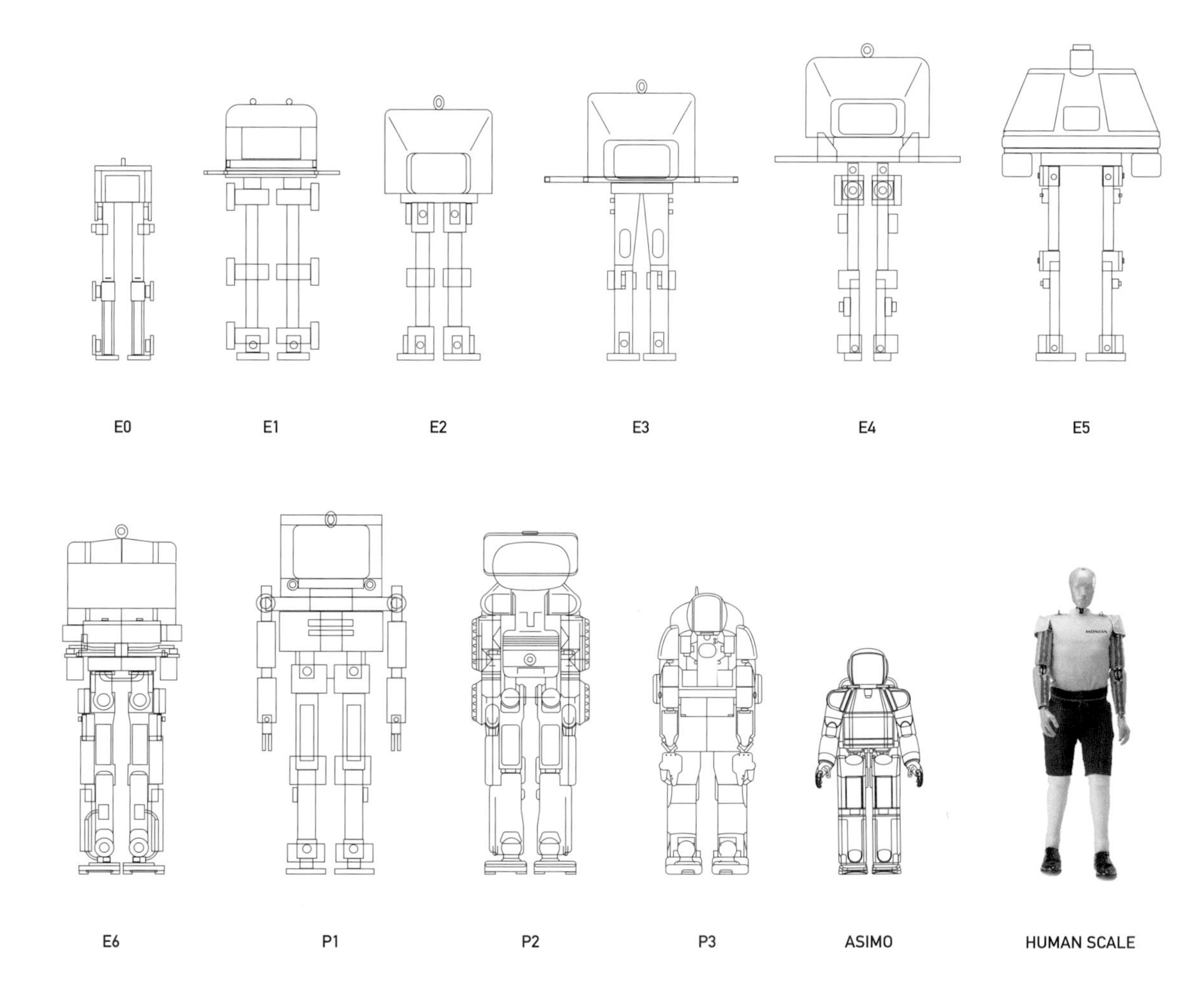
E0
E1
E2
E3
E4
E5
E6
P1
P2
P3
ASIMO
HUMAN SCALE

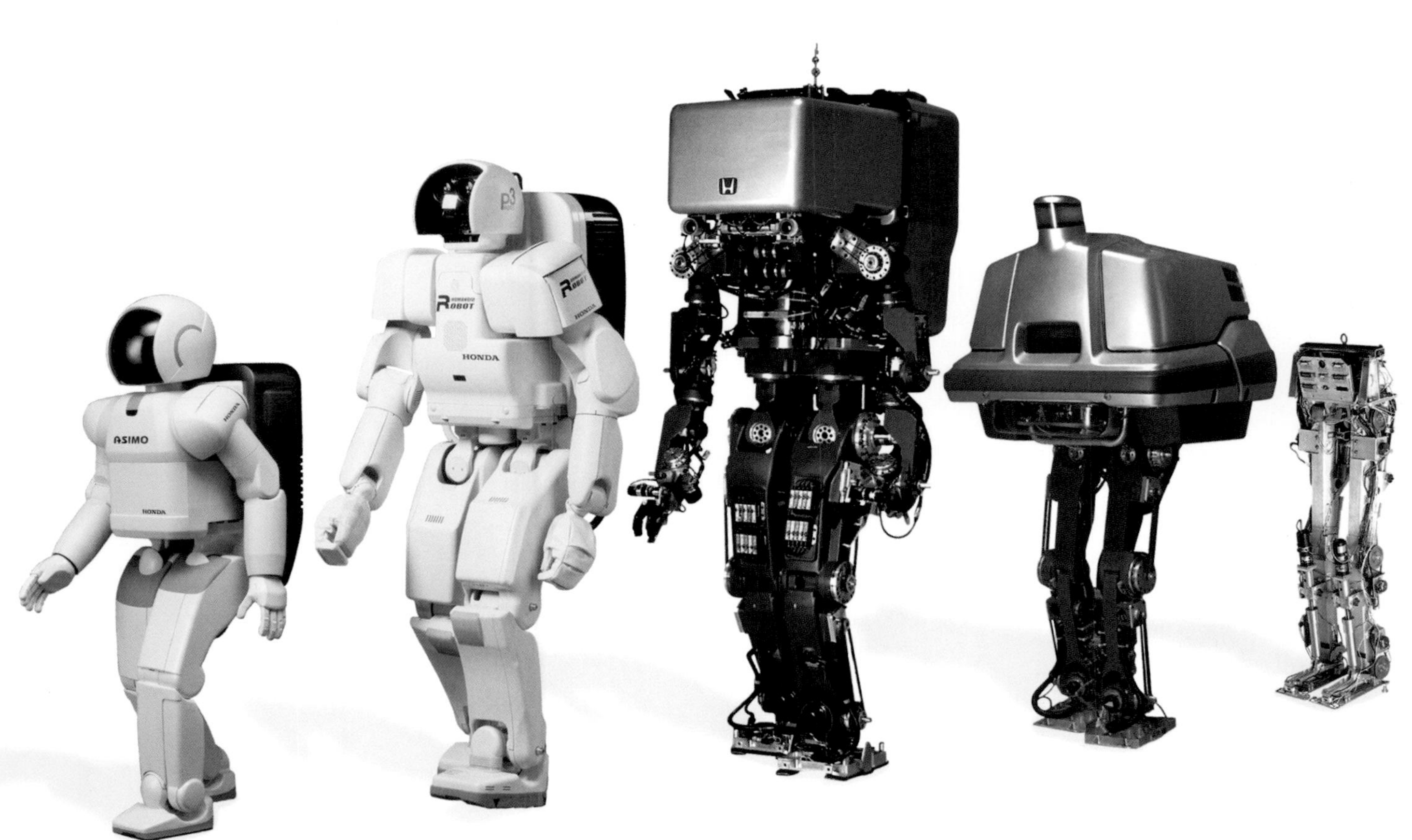
ASIMO
HONDA

doors, and picking up and carrying objects. But it was in 1996 that the world's first self-regulating, two-legged humanoid robot, the P2, was created. This 1.8-metre (6-feet), 210-kilogram (80-pound) behemoth had integral computers, cameras, drives, batteries, radio and other components enabling it to be free from cables and external power sources and independent in operation. This exciting development was reduced in size (to 1.6 metres/5 feet) and weight (to 130 kilograms/50 pounds) in the next iteration, the P3, in 1997. The huge amount of data, research material, man-hours, experimental models and prototypes generated were then brought together for the project ASIMO. For the first time, it was possible to see a future where servile, friendly and functional machines could live harmoniously with their human creators.

ASIMO is a compact (1.2-metre/4-feet), light (52-kilogram/20-pound), intelligent, fully autonomous and fully mobile humanoid for the living and working environment. Using its integral sensors, gyroscope, servos, power supply and computers, ASIMO can achieve flexible real-time walking, change direction and react with stability to sudden environmental movements or obstacles. Called 'i-walk', ASIMO's predictive movement control enables it to turn corners in the same way that a human intuitively shifts his or her gravity towards the inside of a corner, by predicting this displacement of gravity and adjusting its torso accordingly. The result is continuous movement without pauses that is extremely impressive to watch. Smooth and agile movement is the key to the acceptance of robots as more than just automatons.

With total functionality in its torso, arms and hands, ASIMO can simulate the very subtle human body language that is crucial to effective interaction, without misinterpretation, offence or any perception of threat. It is controlled by either a workstation or a laptop via a wireless link, and can be programmed to carry out complex instructions, allowing for various potential applications in such roles as night-time security, greeting and accompanying guests at company offices and museums (the IBM Museum in Japan uses ASIMO robots to greet visitors), as well as operating lifts and other simple machinery.

The scale of ASIMO was developed intentionally to create a feeling of a small, benign presence around the home and office, but one with enough height to reach light switches and door knobs, to work at tables and desks, and to be at eye level when a human is seated. Its styling is a subtle but simple expression of its function, in so far as the way the technology packed into its diminutive and pared-down form intentionally reflects the gentle, non-threatening nature of ASIMO's personality. While ASIMO is not designed for a particular existing market (there is no consumer market for humanoid robots at the moment), its design and styling features many consumer-driven emotional triggers. Its friendly astronaut appearance is reinforced by the visor on its head which can be retracted to reveal oversized, almost comic eyes (which in reality are cameras), and, strangely, are one of its most expressive anthropomorphic details. Watching ASIMO in action is a surprisingly powerful, almost emotional experience, and its successful mastery of human movement makes one begin to believe in the dream.

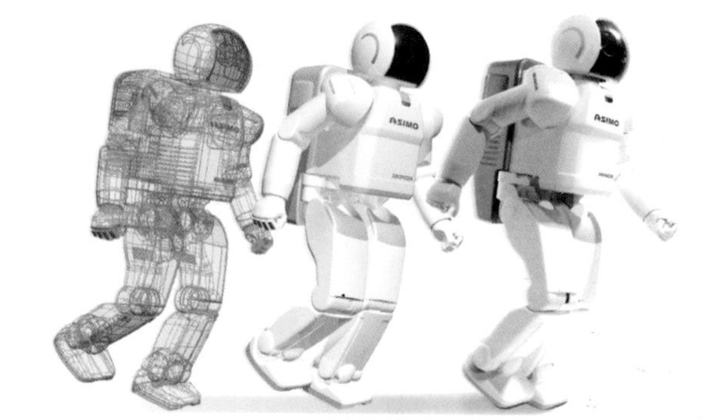

right ASIMO CAD views
opposite Honda CEO and President Hiroyuki Yoshino with ASIMO, who stated 'One never tires of looking at a dream.'

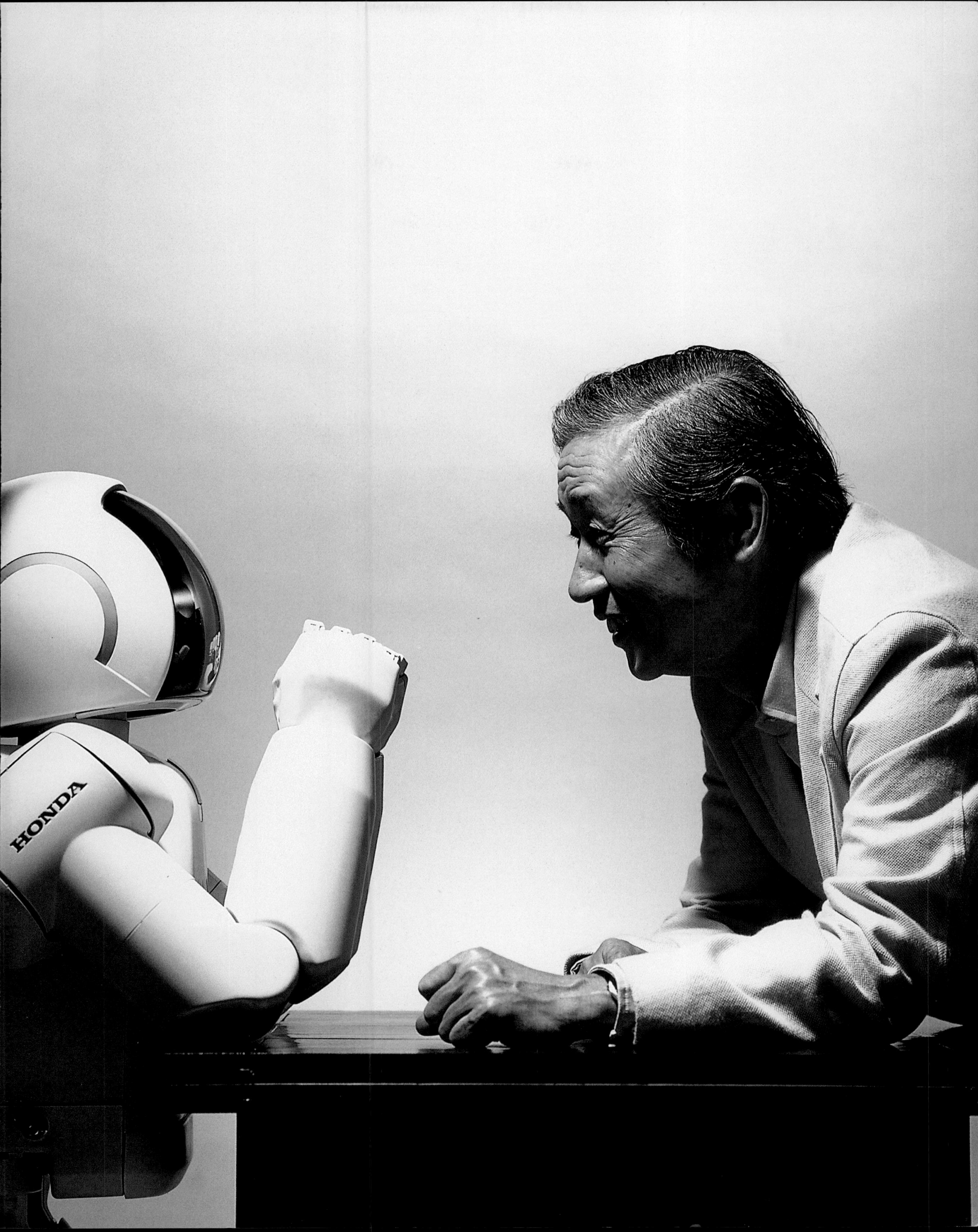
HONDA

Sony Corporation

Aibo (Artificial Intelligence roBOt)

aibo in Japanese means friend or buddy while *ai* means love

ERS-110, ERS-210, ERS-220, ERS-311 Macaron, ERS-312 Latte

SDR-3X and SDR-4X

biped entertainment robots

Q-taro (Quasi-stable Travelling Action RObot)

'healing creature'

Launched on the internet on 11 May 1999, Sony Corporation's first production run of 5,000 Aibo ERS-110 robotic pets was completely oversubscribed. The 3,000 units earmarked for the Japanese market sold out in just 20 minutes. A second production run was set up in November and a further 10,000 produced, but incredibly, Sony received more than 130,000 orders for its ¥250,000 (US$2,500/£1,800) cyber pet. The Aibo frenzy took Sony completely by surprise as internal estimates had predicted initial sales of only 2,500.

Aibo does not pretend to be useful around the home, but rather plays on our emotional, somewhat illogical capacity to adopt and befriend technology whether it is dressed up as a lion cub or a dog or anything else. Aibo has a basic and uncomplicated range of six emotions - happiness, sadness, anger, surprise, fear and dislike - and a set of four desires - a yearning to love, the urge to explore, the desire to move and the need to be recharged. Mix these ingredients together and you get an artificial emotional response roughly equivalent to real life.

Yet all this is child's play when compared with 'medium future' technology, which inevitably will make Aibo's limited emotional capabilities seem very ordinary. In some ways, products in this sector could be characterised and measured in terms of 'emotional potential' rather than in 'processing power', as is

opposite Sony Aibo ERS-220

currently the case in personal computing. Ironically, Sony predict that medium to long term, 'thinking' robots will surpass PCs as essential domestic technology.

Aibo ERS-220

Sony's Aibo has made a speedy transition from prototype to leading-edge contemporary robot-as-best-friend. There have been several iterations of the product since its commercial launch in late 1999, with the first- and second-generation models (ERS-110 and ERS-210) styled rather like a baby lion cub, in the early 1990s. In fact the Aibo lineage goes way back to the 60s when Sony first began researching artificial intelligence. Although the current Sony Aibo kennel now holds the cuddly and cute Macaron and Latte puppies (ERS-311 and 312), which are targeted at the younger robo-fan, the very latest top dog is the ERS-220, a model which defines the genre in all its technological finery. This version has many new tricks, like its own 75-word vocabulary, a camera for taking its own JPEG photographs and, of course, the crucial interactivity that allows its owner to raise it from puppy to hound, influencing its moods and reactions to the human environment, all from a wireless (LAN) link between a laptop and the lap dog.

Launched in November 2001, the Aibo ERS-220 is much more a futurist German Shepherd than a soft spaniel. It uses LEDs around its head to communicate its moods to its owner, while a retractable headlight is used to express itself more forcefully. In response, the owner is able to pet the cyber-dog via touch sensors in the head, back and tail. Interestingly, the Aibo doesn't have a reset button, so if you mistreat your pet or want to restart it from scratch, you have to buy some special software. It pays to look after your Aibo well from the outset.

The Aibo ERS-220 has also been designed to counter that common human response to technology - impatience. The standard way to bring up an Aibo is to educate and bond with it, taking it through a series of lessons and life stages. But for those who find that process tiresome and time-consuming, there is an Explorer software version, which allows owners to skip the puberty part of ownership and experience Aibo at full maturity without putting in the hours. They can still influence their 220's personality, but without the husbandry.

Conversely, owners who wish to can use some software called 'Aibo Life 2', which requires them to raise and interact with their Aibo from the moment it is switched on, using its touch sensors, or 'Aibo Choreography'. This gives the owner increased opportunities to teach the ERS-220 original reactions, movements and tricks using voice commands.

below early Aibo prototypes
opposite first-generation Aibo ERS-110

The somewhat aggressive styling of the 220 (when compared to previous versions at least) does represent Sony's desire to move the product away from its image as a toy towards a more serious appeal for the older, techno-savvy consumer. Whereas the original Aibos were unashamedly cuddlier, friendlier and less 'intelligent', operating on a more basic level of human response, the ERS-220 elicits emotional responses in humans that are closer to those found between people and cars, or other technology-based products, whilst retaining the element of wonder so crucial to this sector.

As always, life is beginning to imitate art. In William Gibson's book *Neuromancer* there exists a world where man and machine are physically and mentally entwined, the distinctions between human and artificial intelligence blurred. It is a celebration of a sophisticated interactive technological utopia. Conceptions such as these have existed in our imaginations for decades, and one cannot easily detach them from the world of more mundane, everyday consumer products. The realisation that fiction still races ahead of reality does not necessarily prevent us from perceiving a robotic dog

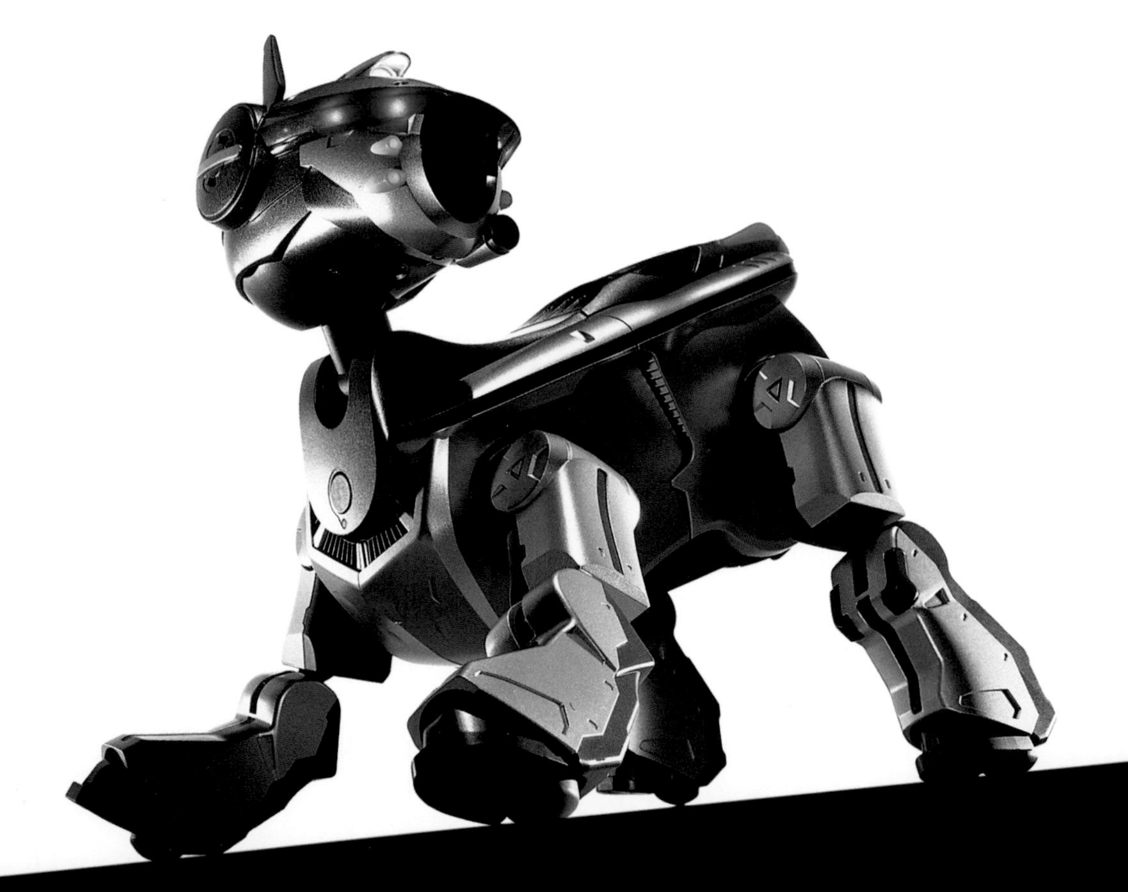

as a magical addition to the family, and, as a consequence, still feeling disappointment at the level of electronic emotion or behavioural interaction that is possible. As in Gibson's world, we still need to suspend reality and engage our imaginations, or at least until familiarity and boredom set in, at which point most products are discarded anyway.

Hence the Aibo's learning curve. One can teach Aibo, over time, to carry out very simple tasks, but the control factor is key to our on-going interaction with robots and creates a feeling of ownership and need. Naturally, a cultural phenomenon like Aibo precipitates a range of associated products, and there is now a thriving Aibo industry producing websites, shows and meetings, books and magazines to connect and inform the legions of fans and owners all over the world.

opposite (clockwise from top) second-generation Aibo ERS-210, magazines for Aibo enthusiasts and Aibo ERS-220
left Aibo ERS-220

Aibo ERS-311, ERS-312

With saccharin names like Macaron and Latte, it's plain to see how Sony perceive another possible generation of Aibo fans. These cuddly characters represent a fascinating juxtaposition of cartoon characterisation with a cool technological heart. As with other Aibo cyber-pets, these artificial pups have a plethora of features to enable the owner to influence how they develop, behave and react to certain environmental stimuli. They include a 75-word vocabulary, a camera and a motion detector, but in some ways their most interesting feature is their differing temperaments. Macaron is the more mischievous of the two and Latte the more compliant. Their personalities are communicated by a combination of body language and visual indication. This cartoon-derived behaviour (head to one side with ear cocked to suggest listening, etc.) means that a basic, but endearing understanding between owner and Aibo puppy is possible.

Since the introduction of arguably the first-ever cyber-pet, Bandai's Tamagotchi in the mid-1990s, the strange and particularly Japanese concept of developing an emotional tie with an inanimate object has become almost normal. This product development challenged our natural instinct to resent the creeping encroachment of technology in our lives, and its enormous success, in Japan especially, has encouraged manufacturers to emphasise the 'emotional' potential of similar new products and services.

Yet the implications of Aibo stretch far beyond its current value as an indulgent plaything. If one remembers the growing impact of an ageing population, particularly in Japan, Aibo and its contemporaries represent the first step towards a product that will eventually offer a completely new set of relationship values between products and people. These values are not based solely on form or function, but on a connection with the hearts and minds of consumers, especially the lonely or incapacitated, for whom the comforting emotional aspects of these robots are most important. The only restriction in this exciting field is the imagination of the engineers and designers. Sony's Macaron and Latte are just the beginning of something huge.

below, opposite and overleaf
Aibo ERS-312 Latte (left) and ERS-311 Macaron (right)

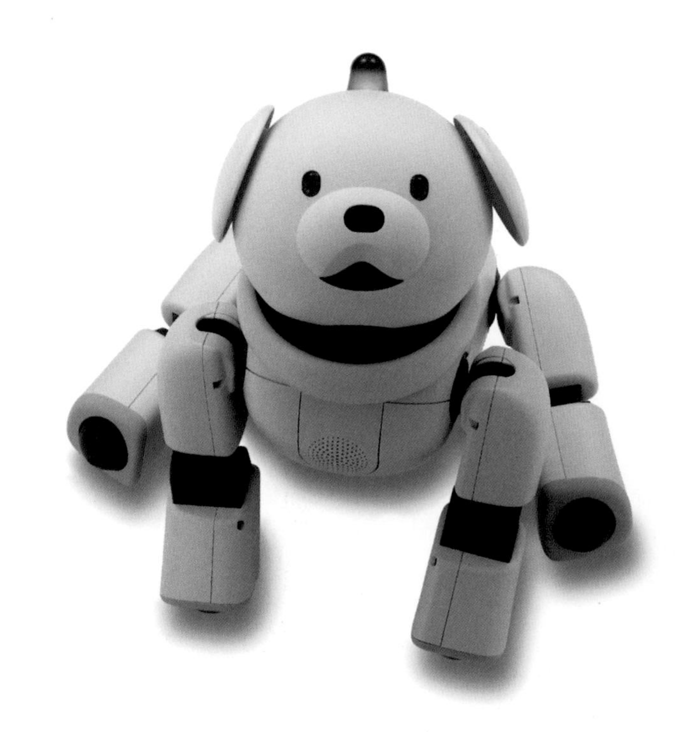

Sony SDR-3X and 4X biped robots

The Aibo programme is just the start of Sony's plans for entertainment robotics. 'Entertainment' is Sony's key word here, expressing the main difference between Sony and, say, Honda, one of the very few companies seriously researching into and building intelligent companions for humans.

The SDR-3X is a compact, 500-mm (20-inch) high bipedal robot launched in 2000 to an enthusiastic public already familiar with the cybernetic charms of the Aibo and Honda ASIMO programmes. The appearance of the SDR-3X reflects the Sony engineers' desire to replicate many of the features of human mobility and emotion rather than the appearance and purpose of a pet.

The first bipedal robotic machine, Wabot 1, was created in the early 1970s at Waseda University in Tokyo and served to highlight the problems associated with simulating the human ability to walk upright through obstacles and up stairs. Both Sony and Honda decided to research and develop studies into this problematic area, each pursuing different approaches. Honda was the first to reveal a walking and climbing android, the P3, followed by the more advanced ASIMO, with the even greater mobility needed to climb stairs.

The Sony SDR-3X is diminutive when compared with the 1.2-metre (4-inch) tall Honda ASIMO. But whereas ASIMO was developed strictly to undertake extreme, tedious or unpleasant tasks, the SDR-3X is an advanced performance-orientated entertainment robot with total body control via advanced control architecture similar to Aibo's (Open-R). It is still within the bounds of what we expect a robot to look like – about 50 per cent human, 40 per cent machine and 10 per cent science fiction, and as with so many Sony products, the reduction of scale and the expansion of function were key objectives. The SDR-3X has integral dual 64-bit RISC processors which are fed from an on-board CCD camera that enables the robot to predict obstacles and to act accordingly. Stable walking is at the mercy of physics.
To counteract and react to the forces of inertia and body weight, and to synchronise movement, the SDR-3X combines posture, touch sensors and data from the CCD camera. Thanks to its 24-degrees freedom of movement and a walking pace of 15 metres (45 feet) per minute at 600mm (2 feet) per stride, the SDR-3X can kick a football. It can also recognise faces and speech, and respond verbally.

Two years after the launch of the SDR-3X, Sony announced its successor, the SDR-4X. A 580-mm (2-feet) tall development of its bipedal predecessor, with 20 per cent better performance and enhanced abilities, this version appeared at Robodex 2002 in Yokohama. Its technology is directed in four key areas – mobility, voice and visual recognition, and interaction with humans. Mobility is achieved by a 'real-time, integrated, adaptive motion-control system', which, put simply, is the smooth, real-time control of all 38 joints of the biped via the data collected by its on-board sensors.
This control system enables the SDR-4X to cope with uneven or angled surfaces, navigate obstacles, change direction, and even to resist being pushed over. The twin CCD cameras in the head of the biped are the primary sensors used to identify objects and obstacles. These work together processing the parallax or converging data. This works in a similar way to the twin beams of light used in the famous

this page SDR-3X biped robot

Dambusters raids of the Second World War. The short- and long-term memory of the 4X also means that it can memorise and avoid things such as walls or stairs, while reacting and adapting to any new or sudden obstructions.

Seven on-board microphones are used for voice recognition, to identify and interact with humans. An individual voice is recognised as tonal data which, when combined with the 4X's ability to distinguish facial features, even against complex backgrounds, means that the robot can communicate with up to ten different people. When linked to a personal computer via a wireless LAN, the biped can recognise words and speech patterns and even learn new words not contained in its on-board dictionary. Synthesised speech features mean that it can talk and even produce a singing voice, complete with vibratos, when music and lyric data are inputted via a PC. When twinned with the SDR Motion Creator, which enables the biped to create personalised movement performances, the SDR-4X effectively becomes the first all-singing, all-dancing robot.

The product design is careful to avoid potential finger traps in the biped's joints, important when one considers the close interaction with humans, and the styling provides an unthreatening but distinctive appearance, essential to domestic acceptance, without resorting to robotic cliché. Expressions are enhanced by the addition of four axis movements in the head and wrist, plus the five individually moving fingers in the hands.

left SDR-4X biped robot

Although not available to buy, the SDR-4X is a significant embodiment of Sony's belief in the simulation and enrichment of human life. It may even represent a departure from a traditional stylised robotic hero towards a real and controllable replacement in the hearts of consumers and around the home.

Sony Q-taro

The Q-taro prototype (meaning Quasi-stable Travelling and Action RObot) represents a unique developmental direction in the communication between humans and robots. This spherical 'healing creature' is a robot only in the sense that it is controlled by, and responds to, human interaction, but in far more subtle ways than its cousins in the Sony robot family. In the same way as petting a domestic cat can reduce stress and lower one's heart rate, Q-taro is designed to be a restrained and undemanding presence about the home, creating a warm and anxiety-free alternative to conventional robots.

Among its contemporaries designed to inhabit a human environment, the 1-kilogram (2.25-pound), 170-mm (6.6-inch) diameter Q-taro sphere is an enigmatic product both conceptually and functionally. It is energised only when it senses the presence of a human, something it does by monitoring body temperature. Rolling across a surface autonomously, it can overcome obstacles and automatically draw itself towards its own charger when recharging is required. It expresses its moods throughout the day by adjusting the brightness of its body and through its movements. It also responds to speech through voice recognition technologies and will react to music when connected to a music source. The emotional 'maturity' of the Q-taro is evident by the way the sphere learns about and adapts to its environment, continuously expanding its repertoire of reactions and functions. Future Q-taro robots will be able to communicate with each other, providing for the exchange of all kinds of information and experience between the units and their owners.

Although Q-taro is not available commercially, it does signify how a subtle iteration of robotic technology can be as interesting and influential as the more obvious examples. It will also, hopefully, lead to a wider acceptance of robots about the home.

opposite Q-taro 'healing creature'

AUFLADEN
SONY

NEC Corporation

PaPeRo (Partner-type Personal Robot)

In 1997, an NEC team of mechanical engineers, computer scientists, specialists in voice-recognition technology, interaction and industrial designers started out on a journey into artificial intelligence.

The aim was to develop a project that would reflect NEC's thinking and technological virtuosity, and which could become a media-friendly communication vehicle for the corporation's ambitions. This kind of project, where billions of Yen are pumped into the pursuit of a company icon, is only possible where the organisation in question already has a huge R&D budget culture and where long-term objectives are as important as short-term production and development.

The overarching goal was to create a robot capable of autonomous motion and bring together technologies from different areas of research within NEC. Technologies like voice and visual recognition were merged with 'mobile architecture' to achieve this. Mechanical electronics, actuators, sensors and, of course, computing power, were developed along with intercommunication technologies like Bluetooth and LAN connectivity.

One of the main questions was how this iteration of artificial intelligence should be introduced to intended users. Should it be presented and used as an electronic pet, or as an assistant, a partner, or maybe just as domestic equipment? This question inevitably led NEC's team into researching the role that pets play in the home environment and attendant notions of harmony, companionship and playfulness.

Eventually, however, a bigger issue became clear. Whilst we accept and understand a certain level of misbehaviour from an organic pet, for an electronic one to fail to live up to expectations is a more serious problem. Of course, there is a gap between the promise of technology and the reality of what it can actually deliver, and this occurs in many areas of modern life. We all experience disappointment when technology fails to do quite what we had hoped, but it is a particularly acute problem with an electronic assistant or

opposite transparent NEC PaPeRo showing internal components

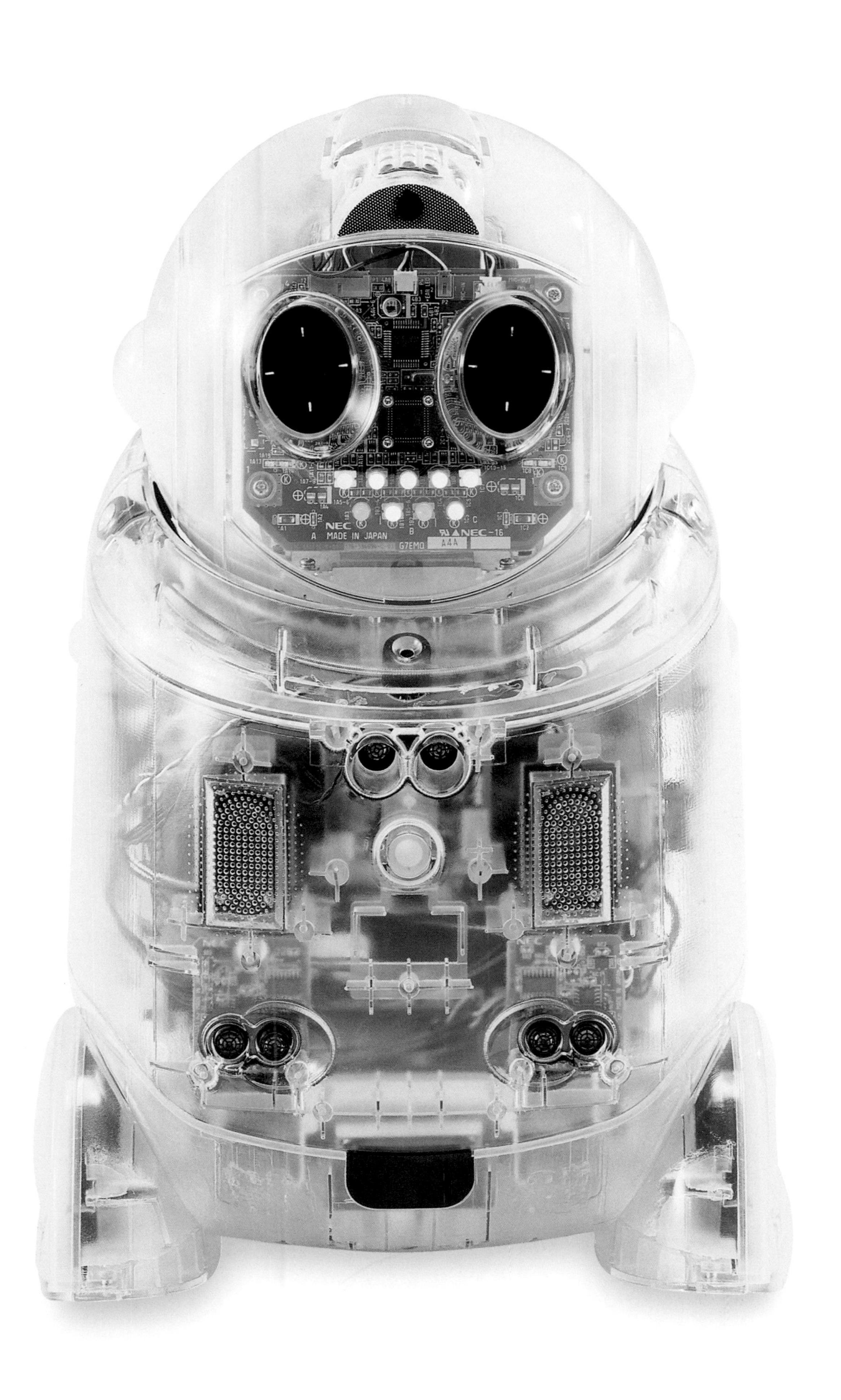
NEC
A
MADE IN JAPAN
B
C
NEC-16
G7EMO
A4A

robot. Disappointment can quickly lead to anger with something as emotionally invested as a personal robot if it does not fulfil its promise or potential.

From a functional point of view, the key technologies for the PaPeRo were picture/vision and speech/voice recognition. It was developed to use as few actuators as possible while still being able to animate the recognisable gestures and emotions crucial for interaction with its owner, and remain mobile. Vision and picture recognition is achieved by twin CCD cameras, and the speech and voice recognition function uses a specially developed NEC system called SmartVoice, and a custom built MMI (Multimedia interface) devised to give PaPeRo a high level of usability.

When PaPeRo meets a family member, it will attempt to recognise them by referencing its memory of previous encounters and gradually adapt its behaviour and attitude to them, depending on those previous interactions. The designers intentionally developed PaPeRo's personality to be inquisitive and helpful, always asking questions about favourite things or interests and remembering the responses; and communicating useful information about news or the weather downloaded from the internet.

Apart from carrying out the simple functional tasks of domestic life, like changing television channels and switching on or off light, air conditioning and entertainment equipment, PaPeRo can also send and receive video e-mails and operate as a security system, using its motion sensors and image-recognition technology to detect intruders.

this page and opposite PaPeRos

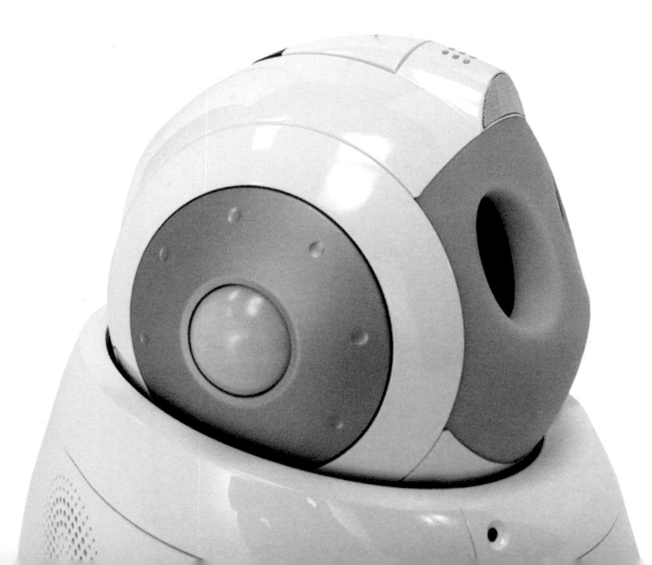

The project gained momentum just as Honda and Sony pursued their own objectives within the robotics industry and the concept of an artificial partner for the home grew. The technology involved in this project is regarded by most of these corporations as 'middle future' (while that used in products to be available in a year or two is 'near future'). True artificial intelligence, or thinking machines, is described as 'long future'.

Product designers were involved in the PaPeRo project from the outset, including the initial discussions about developmental objectives and research into behavioural patterns between humans and their pets. This was crucially important for the team, as the designers used a prototype-driven process that required constant conceptual objectivity. It was felt that for users to interact well with PaPeRo, the robot had to have a definable character, and once again, key questions about the nature of its 'personality' - cute and cuddly friend or serious technological assistant - became central; as did discussion about acceptable levels of performance in a technology that promised so much. Visual as well as voice recognition is accurate to around 80-90 per cent, but even this high degree of accuracy can bring frustration and disappointment in everyday interactions.

The early design appearance was close to the designers' wishes in terms of scale, character and appearance, an important achievement as this kind of project is so frequently driven by technical considerations rather than the emotional and psychological ones. For instance, PaPeRo's face was a crucial area for development, with many versions produced for comparison to find the right balance of cuteness and coolness while maintaining technical performance on issues such as angle of sight.

PaPeRo is referred to as 'he' rather than 'it', and generally perceived to be a boy (rather like a reborn, reshaped Astroboy for a new generation). Even so, NEC's designers have tried to produce an intelligent, likeable robotic character rather than a simulated human. If PaPeRo wedges himself in a corner or falls over he will call for help and if none arrives he will fall asleep. PaPeRo wants to be part of the family unit, to be a friend, and always asks for information as well as providing it.

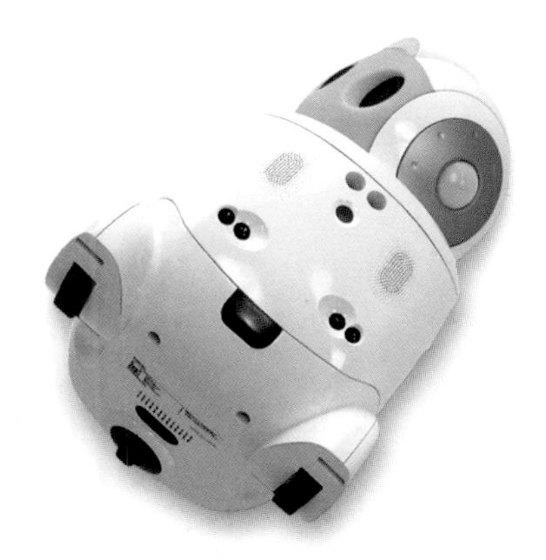

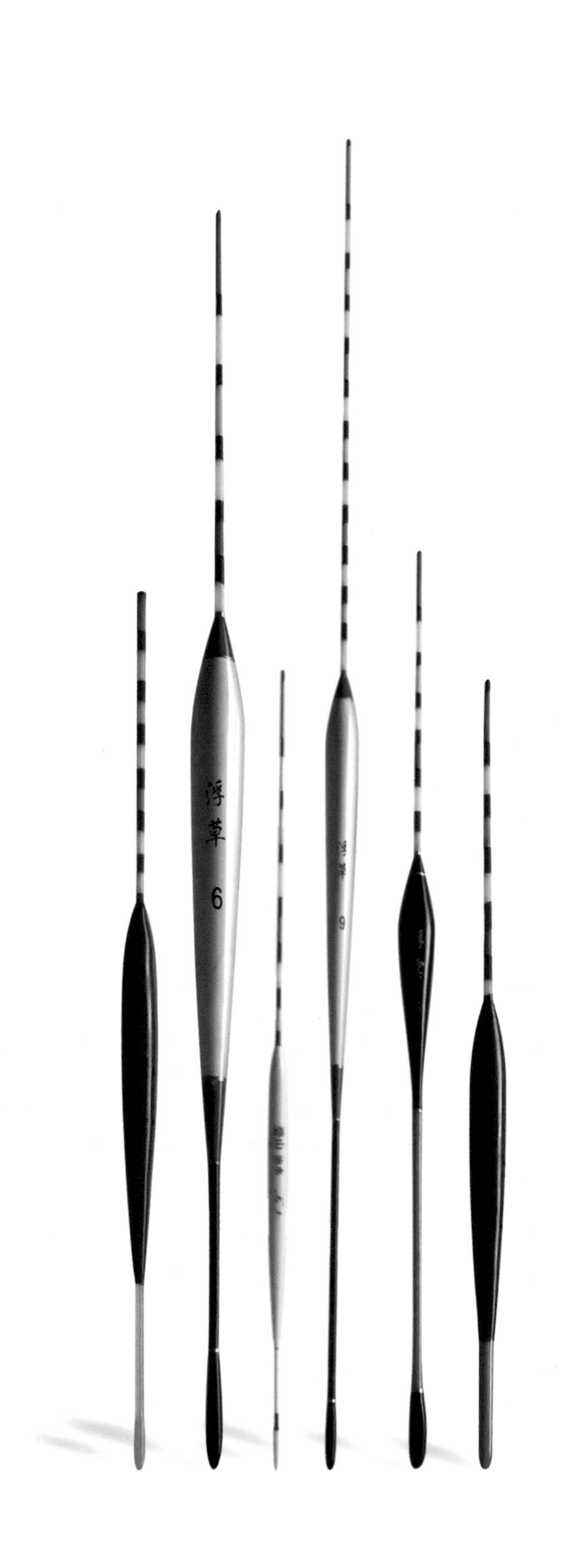
浮草
6
9

2

perfecting perfection

CONTINUOUS PRODUCT DEVELOPMENT

This chapter looks at examples of the constant striving in Japanese industry to hone and develop technologies. The Japan Railways Shinkansen (or Bullet Train) was introduced in 1964 and remains the epitome of fast train-travel excellence through a continual development process in which timekeeping and speed are something of an obsession. The Shinkansen is almost a 'robot hero' in a Japanese sense, occupying both a functional and emotional place in modern Japan. This section will trace the origins of the service through to current designs and future developments, including the high-speed Maglev (magnetic levitation) trains.

Other examples of this process of continuous product improvement that will be featured in the chapter include Shimano's 7700 bicycle wheels – universally acknowledged as among the finest in the world – and as much at home in the 2002 Tour de France as on a bike bought in the high street – and the Stella fishing reel from the same company. This fishing reel is at the forefront of current material and functional performance and its design celebrates the contemplative nature of angling. Other examples include Bridgestone's Potenza tyres which, utilised by racing partner Ferrari, are currently dominating Formula One racing. The Pentax 645NII camera is the result of over 80 years of camera and lens development at one of Japan's most celebrated camera companies, Asahi Optical, while Mizuno's Wave Cup football boots, worn by Rivaldo, certainly contributed to Brazil's 2002 World Cup victory, ironically won in Japan. With Yamaha's SFF-2 archery bow and NEC's SX-4 Supercomputer, these products represent the sheer diversity of near perfection in Japan.

'Perfecting Perfection' is about the development of technology through a patient, logical, rational process. This is achieved through committed teamwork, collaboration and moments of genius.

opposite traditional fishing floats

Yamaha Corporation

Super Feel Forged 2 (SFF-2) archery bow

There is a neat logic to Yamaha producing sports equipment as well as musical instruments. Musical expression and archery feature heavily in Japanese history and there is a natural, if unfortunate, symmetry between instruments of beauty and enjoyment and instruments of accuracy and, traditionally at least, of death.

Throughout history the bow and arrow have been a familiar part of warfare. Traditionally, whatever they are made from, whether composites of wood or bamboo or other organic materials, all bows have as a common element the vibration caused by the sudden release of energy when an arrow is fired. How this energy is dispersed determines how comfortable the bow is to use, and how accurate the shot.

Modern bows use a stabiliser to help channel these vibrations away from the archer and achieve a more accurate projectile. The direction, flight and velocity of an arrow are governed entirely by the archer at the point of dispatch, and a good bow is required to be controllable, balanced and ergonomic. This is crucial, as the generation of loads on a 1.78-metre (70-inch) bow requires up to 130 kilograms (50 pounds) of pull, and only 70 per cent of that energy is transferred to the velocity of the arrow.

The Yamaha SFF-2 is no bent wooden longbow, but rather a leading-edge piece of sports equipment that belongs firmly in the contemporary technology camp. This modern bow has a light, rigid forged-aluminium handle (forging has a much-vaunted reputation in traditional Japanese sword-making), and a ceramic/carbon limb (the bow itself) which combine to give it such power and accuracy.

The SFF-2 is a very distant cousin of archery bows from the days of warfare, both in terms of material and purpose. But the laws of physics remain and the physical problems of the past are still as relevant today, a true test of strength and accuracy.

opposite Yamaha SFF-2 archery bow

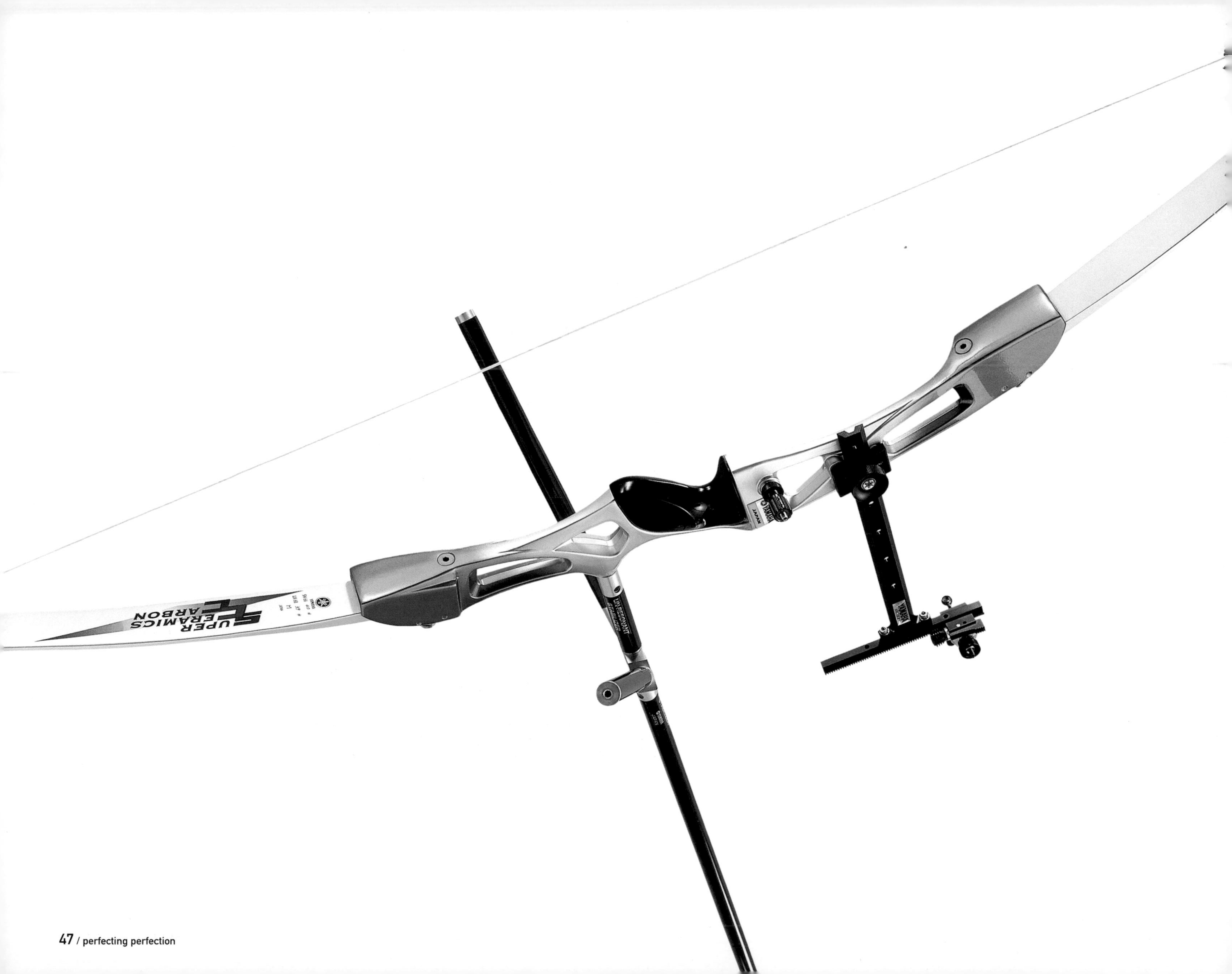
YAMAHA
JAPAN
UN-RESONANT
STABILIZER
UPER
ERAMICS
ARBON

NEC Corporation

SX-4 Supercomputer

In very high-end computing, units are used to measure the number of arithmetical operations that a computer can perform in a second. They are called FLOPS (Floating point instructions Per Second). For instance, 10 FLOPS indicates ten arithmetical operations per second. A GigaFLOPS (GFLOPS) means one billion floating points (or operations) per second, and a TeraFLOPS (TFLOPS) means 1,000 billion floating points per second.

NEC's SX-4 is a Scalable Parallel Vector Supercomputer which scales from 1 GigaFLOPS to 1 TeraFLOPS. An instrument of torture for numbers, it was created for the crunching of mathematical conundrums at extraordinary speeds. The SX-4 takes 40 seconds to solve a complex equation that would take the fastest PC available today 24 hours. It can also simulate a nuclear test and, as a consequence, is arguably key to the preservation of world peace and a safe and clean environment. This kind of machine takes us closer than ever to the artificial 'brain' or man-made intelligence of films like *2001: A Space Odyssey*. For most of us they are an invisible presence, an unseen representation of raw power with a brooding intensity that contributes to the myth of the monster computer in fiction.

NEC's reputation for building the world's most powerful supercomputers began with the SX-2 in 1983, the first computer to work at the Giga level, followed by the SX-3 of 1989, which reached the 10 Giga level. These are computers for the rarefied air of mega-corporations and national defence-level calculations, and it is most unlikely that they will ever be seen outside the establishments or facilities in which they are currently installed.

With the SX-4 (Model 32-N3932), however, NEC's designers and engineers worked together to create a new direction in supercomputing that would address the two latent, but important, issues for these monsters of the darkened machine room – the saving of time (incredible processing speed) and security (safe and reliable storage of customers'/users' data).

With the SX-4, NEC took the opportunity to create an entirely unique visual presence that reflected the confidence and capability of its technology, and one that suggested enormous power and performance. The designers faced the two-fold challenge of designing a product that virtually no one would see or touch, but that would nonetheless show the hidden but amazing qualities of the technology within. Just imagine: cooling fans produce the 'breathing' noise of a gargantuan monster, while tens of thousands of cables and wires are packed within like arteries and veins. At the heart is a giant heat sink, glistening like a jewel, and the brain is represented by stacks and stacks of beautiful CPU PCBs (central processing unit, printed circuit boards). So proud were the designers of their creation, that each one left their initials inside the casings. More prosaically, usability was built in. The hardware was developed to be easily transported, assembled and expanded, and individual components fitted into conventional elevators.

The critical issue of access to internal components for expansion or servicing was solved by the distinctive pentagon format. Sightings of this dark, minimalist monster are rare, but most installations that have an SX-4 also have specialist rooms with mezzanines from which its mighty presence and brooding beauty can be appreciated. The pentagon-based crucifix layout expresses security, symmetry and balance, and the heart of the machine – the red veins – suggests the stacks of processing boards within.
The appearance of this symbolic super-computer was governed by the need for the shortest cable layout possible (the shorter the cabling the better the performance), and the requirement for easy access to the thousands of components in this leviathan of the laboratory.

The SX-4, designed in 1993, is one of the iconic statements of the power of the 'thinking machine' of recent times. However, time and development move inexorably onwards and NEC's 2002 version, currently the world's fastest computer, runs at a giddying 35.9 TFLOPS. In 2004, IBM plan to wrest back the mantle of the fastest computer on the planet with the launch of the enigmatically named ASCI Purple, with a running speed of 100 TFLOPS, and an ultimate target of 1000 TFLOPS in the near future. In other words, soon it will even be surpassing the speed of the human brain.

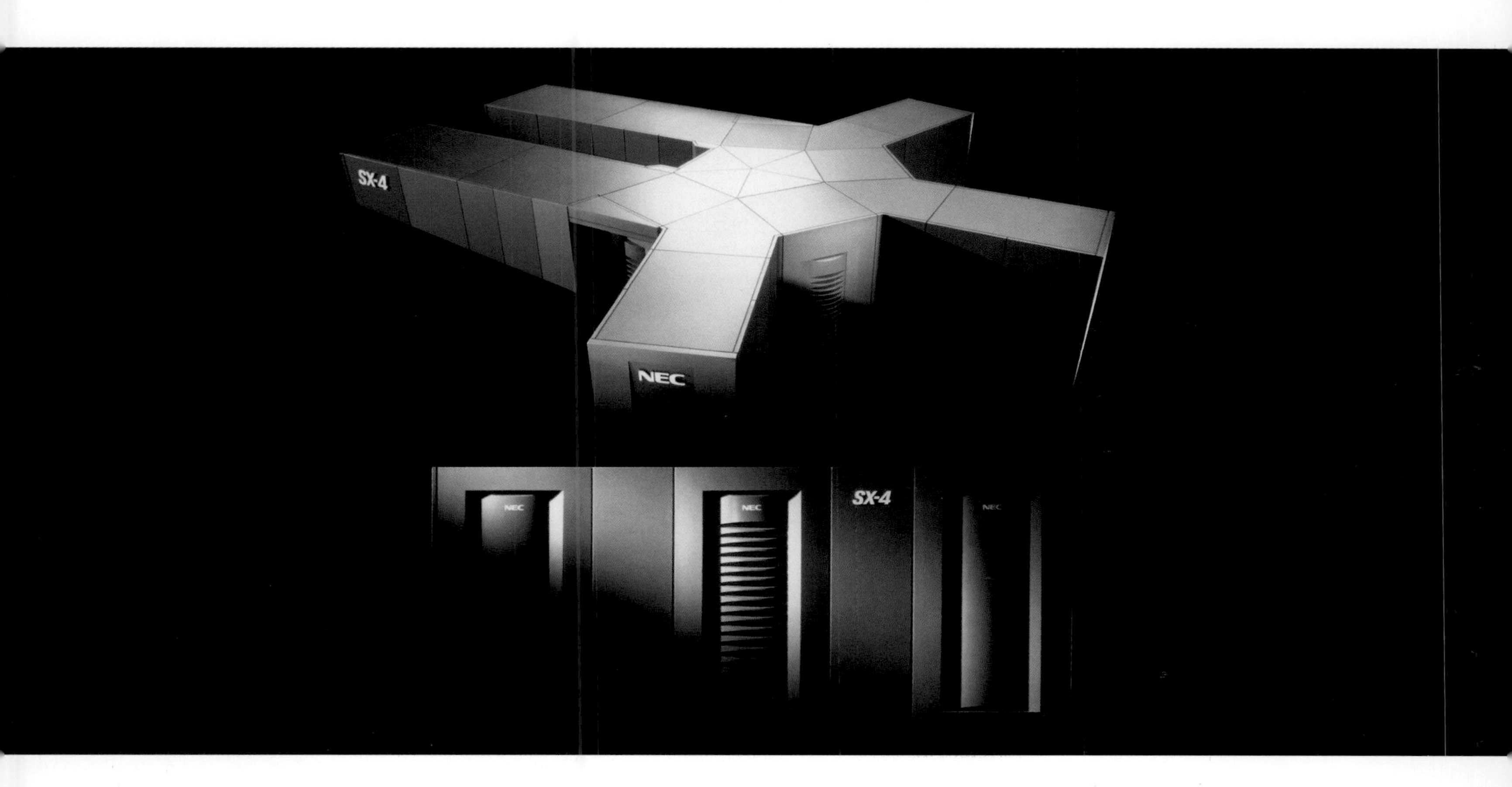

above NEC SX-4 Supercomputer

Shimano Inc.

Stella FW2500S fishing reel

WH-7700 Carbon pro-competition race wheels

Products that are powered by human kinetic energy constitute a sector where we can find sublime examples of Japanese engineering excellence, and where the efficiency and honesty of the product require no embellishment. Whether an archery bow, a musical instrument, a bicycle or a fishing reel, the basic requirements of human interaction are central. While humans can adapt to accommodate shortcomings in product function, this is becoming less and less necessary as products get better and better. This type of continuous evolution is something we find time and again in products from Japan, and two superb examples come from Shimano.

How close the Shimano Stella fishing reel is to attaining optimal efficiency and performance is difficult to tell, but developing this technology much further must be a daunting prospect for the engineers. Designed to express the designers' notion of a new Japanese beauty, the Stella FW2500S was developed as the flagship product of the Shimano range, using magnesium for its strength and lightness and incorporating possibly the smoothest rotation mechanism available anywhere, the Super SHIP (Super High Power system).

This world-class piece of engineering is a smooth, large-diameter master gear, twinned with a floating shaft that achieves a 90.3 per cent transmission efficiency with a masterfully smooth, powerful action. Unencumbered by decoration, the product is outwardly very simple; appropriately so, considering that fishing is about the triangular relationship between man, fish and rod. In either the meditative state of waiting or the animated state that follows a 'bite', the reel is a natural and sensitive extension of the body, especially when twinned with a Scorpion Shaula high-density carbon-fibre rod.

Such reels are making fishing increasingly one-sided, with the odds stacked higher and higher in favour of the angler as a result of the power of human engineering.

opposite Shimano Super SHIP rotation mechanism
overleaf Stella FW2500S fishing reel (left) and Scorpion Shaula high-density carbon-fibre rod (right)

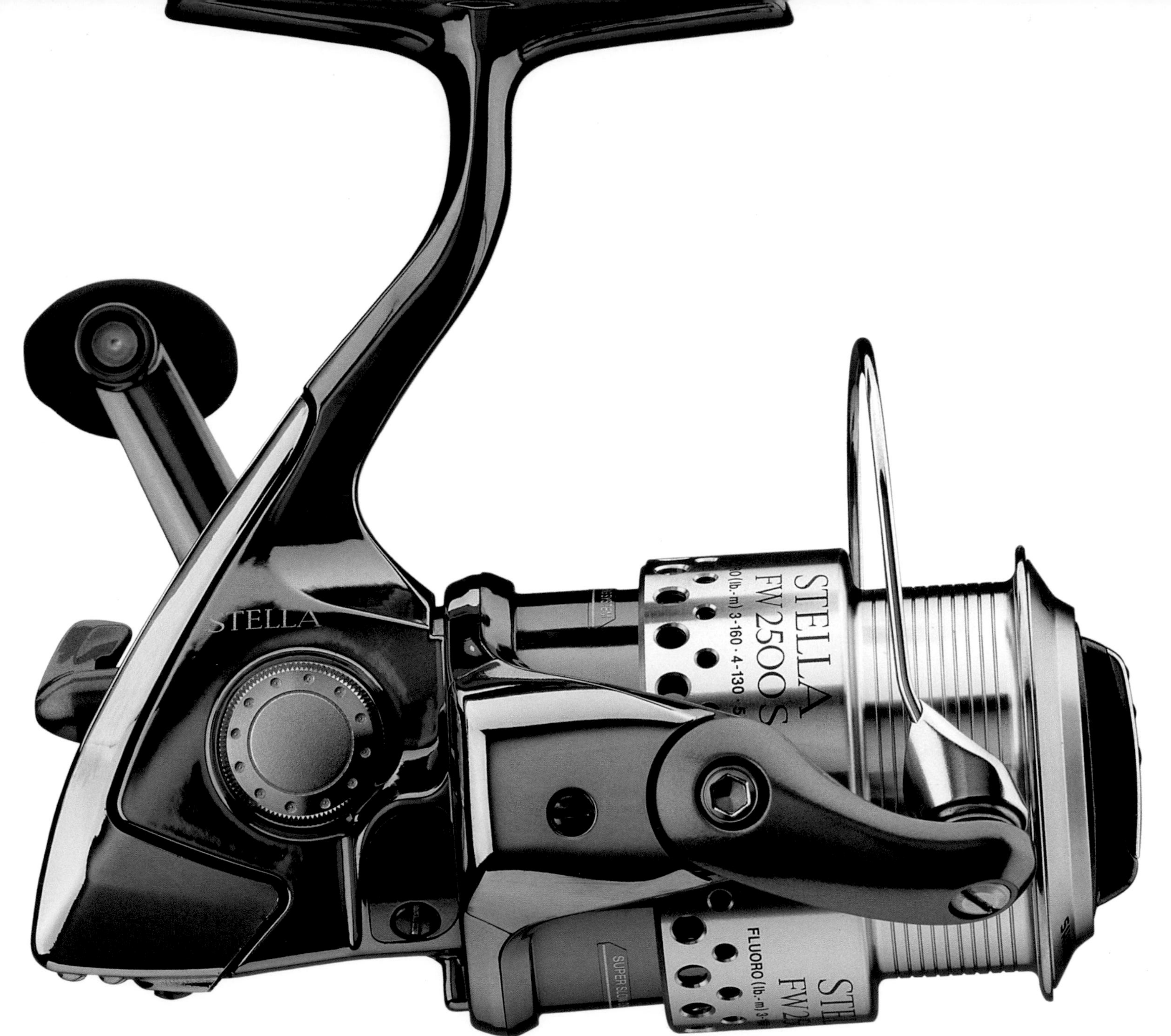
STELLA
FW2500S
(lb.-m) 3-160 · 4-130
FLUORO (lb.-m)
STELLA

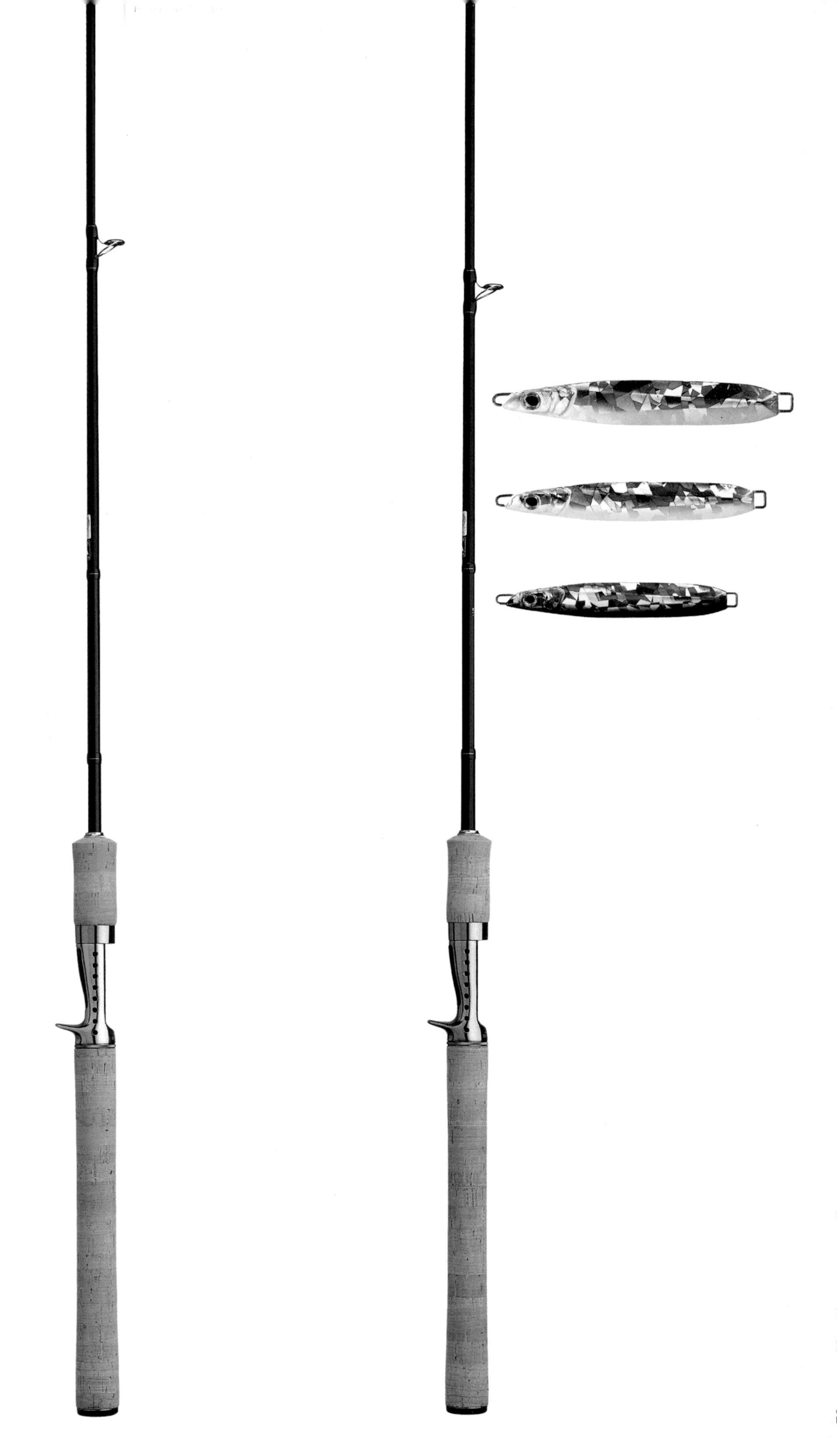

WH-7700 pro-competition race wheels

The family-run company of Shimano has long had a stranglehold on the global bicycle-component market. This is due partly to the sheer scale of its manufacturing arm, but mostly because of its insatiable desire to develop and produce innovative world-class products. Most enthusiasts are familiar with Shimano's brands such as the Tour de France-winning Dura-Ace (road-racing components) and XTR (mountain-bike components), and although the cycling world, like the angling world, is a hive of snobbery and exaggerated performance where equipment is concerned, Shimano is universally acknowledged as setting the standard for bicycle components.

Shimano's history really starts in 1921 with the introduction of the 3.3.3 bicycle freewheel, a benchmark product for the company (up until then, all the best bicycle equipment had been imported from England). Shimano also creates products for two other sectors that have a huge following in Japan: fishing and golf. For Shimano, there are logical parallels between a bicycle and a fishing reel, as the braking and gearing technology is very similar; even golf clubs are perceived as types of gears, with the selection of an individual club for a particular shot resembling the selection of a particular gear on a bicycle for a certain task.

The WH-7700 Carbon is one of the lightest, strongest, most aerodynamic and efficient wheels ever conceived. It is unique in many subtle but crucial areas, notably in the 'lateral crossover' spoke pattern. This innovation means the wheel can support a 90-kilogram (200-pound) rider with just 16 x 14-gauge flat-bladed stainless-steel spokes, which considerably reduces the weight of the wheel. Further weight is saved by anchoring the spokes in the rim sidewall, which gives a very light, aerodynamically-efficient rim and increased rim-to-hub triangulation, which in turn results in increased lateral rigidity. Carbon fibre is used to enhance the weight-to-strength ratio of the wheel rim, which fractionally reduces the amount of effort the rider needs to exert to rotate the wheel. Shimano Dura-Ace hub components are housed in a titanium freehub body for strong, reliable performance. The WH-7700 front wheel weighs just 650 grams (30 ounces).

right Shimano 7700 professional bicycle wheels
opposite 7700 Carbon professional wheels

CARBON
SHIMANO
CARBON
SHIMANO

Asahi Optical Co., Ltd

645 and 645NII cameras

Asahi Pentax is a Japanese company whose past plays a huge part in its current activities and direction. The company has a legacy of optical excellence, going back to 1919, when it was founded as a camera-lens polishing business. After the Second World War, the company founder, Saburo Matsumoto, decided to add camera manufacturing to its expertise, taking advantage of the dearth of entertainment in the immediate post-war period and the already existing popularity of photography in Japan. At the time, most domestic and imported cameras were either twin-lens reflex or 35mm rangefinder models. Matsumoto set out to create a 35mm single-lens reflex camera, a format that he considered to be the next big thing. Based on an old, pre-war 6 x 6-format SLR camera, his engineers produced a prototype 35mm SLR, which was shown to several camera traders. Only one, Hattori & Co. (today's Seiko Corporation), saw its potential and agreed to stock it. This was the Asahiflex I of 1952, and it was the company's and Japan's first-ever mass-produced 35mm camera, and one which is regarded as a significant landmark in camera design.

For *aficionados* of camera design, the Asahiflex I's specification was impressive – a mirror that would remain up while the shutter release button was held down; a screw-type 37mm lens mount; shutter speeds from $^1/_{20}$ to $^1/_{1,500}$ of a second; and a direct optical viewfinder. The Asahiflex I, more than any other model, set the standards for the entire Japanese camera industry, which, of course, was to become the world's leader in 35mm SLR camera production.

Throughout the next 50 years, Pentax's development of the 35mm format was characterised by several landmarks in design and technology – the Asahiflex II of 1954; the 'Pentaprism' and the instant-return mirror in 1957; the TTL (through-the-lens) Aperture-Priority AESystem, which was incorporated into the Asahi Pentax SP of 1964 (the year of the Tokyo Olympics); the bayonet-lens mount on the K2, KX and KM models in 1975; the LX of 1980, which, after over 20 years, is still the company's top non-autofocus camera; and the ME-F of 1981, which was the world's first autofocus TTL SLR camera.

opposite anatomy of a Pentax 645NII

Cameras are one of those kinds of product that develop a personal and emotional value over time, and most of us remember our first camera, just as one remembers one's first watch, computer or car. The camera is a product that rarely wears out, rather, mellowing with age – the scratches and dents become evidence of its history. Many photographers take pride in ownership of cameras that display the marks of experience and adventure, every one representing a story or incident.

For many 30- or 40-year-olds, their first camera might have been the venerable and reliable, yet affordable, K1000 (introduced in 1976), the basic SLR of choice for photography departments in art colleges the world over. This robust manual SLR, with its lens compatibility, high image quality and general usability, remains an essential kit for beginners, and a favourite of professionals. Personally, my first-ever SLR was a younger cousin of the K1000, the MX, which I still own. It remains a treasured part of my creative education; of experimentation and experience.

Today's large-format cameras remain tied to a kind of timewarp style, of which the Asahiflex is representative, where the materials, finishes and ergonomics are engineered rather than designed, where the results are more important than convenience, and where the role of the photographer has as much to do with the alchemy of the process as with the creativity of the image. Large professional formats like 5 x 4 are uncompromisingly functional cameras for photographers who like their equipment to have the appearance of medical or aeronautical gear and where the solidity of the product's construction is matched by the density of knowledge required to use it effectively. This type of photographer is the operator, the technician, and of course, the artist.

The Pentax 645, launched in 1984 as a medium-format field camera and still available today, is an autofocus 6 x 4.5cm SLR camera. Best described as a 'semi format', occupying the space between 35mm and 6 x 9cm format films, it offers photographers the mobility and convenience of a 35mm SLR, with the image quality and control of a larger-format camera.

At best, the 645NII could be described as unconventionally attractive, but its real charm is in its machine aesthetic, where pure functionality leads to a purity of appearance. Its appealing but

below Asahiflex I and Pentax K1000 cameras

brooding ugliness is conditioned by its purpose, with a kind of nobility usually reserved for low-production, hand-built, professional equipment. In fact, the NII's sober design is due to the conservative tastes of professional users of previous 645 versions. The styling was purposely kept restrained and traditional in appearance.

Launched in 1997, the Pentax 654NII, complete with motor wind and viewfinder, is almost as convenient to use as a 35mm camera. The new version has the benefit of the world's first large-format autofocus and user-orientated features include fully customisable functions. What on the surface may seem like a jumble of controls is, in fact, the realisation of years developing an analogue interface, originally developed for the Pentax MZ5 range of 35mm SLR cameras. The speedy and effective use of controls is all about the interpretation of information, and the view adjustment control, drive mode dial, preview lever and condition check system are all visible through the viewfinder.

The 645NII is a handmade camera; Pentax employs just two very skilled and specialist assemblers for this difficult task. There are many complex electronic connections to be made, all by hand. The die-cast aluminium body contributes to the camera's relatively light weight of 1.28 kilograms (45.2 ounces). A unique characteristic of the NII is the special relationship between the analogue and the digital, between the traditional mechanical and electrical components. The electronic provides accuracy and controls the mechanical, which in turn provides durability. There are no plastic parts, only hardened, tempered metal gears and cams. A mirror-lock mechanism, which holds the mirror after the shutter is released, prevents the minute vibration caused by the shutter operation, which can affect the image on the film, particularly when zoom lenses are used.

The Pentax 654NII represents a logical move for those wishing to jump from a 35mm to a larger-format camera, but one which is not larger than many contemporary 35mm SLRs.

left aluminium chassis for the 645NII
overleaf elevations of the Pentax 645NII

PENTAX
smc PENTAX-FA 645 1:2.8 75mm
MADE IN JAPAN

10 15 3 7 ft m
22 16 8 4 4 8 16 22
A 22 16 11 8 5.6 2.8
645NII
ISO
100
OFF ON
M.UP
AE-L
STEP
±0
1000
500
250
125
60
30
15
8
4
2
1
2S
4S
B
X
A

C S AF
AF
AE-L
PENTAX
120
OPEN

MF AF

Mizuno Corporation

Rivaldo Wave Cup football boots

Considering the very personal nature of football boots – representing a physical and psychological extension of the foot – it is not surprising that there were only two key individuals behind the development of Mizuno's radical 'Wave Cup' boot. Rather like the essential collaboration between a director and an actor, the central roles in this project were played by Takeshi Oorei, a project development manager at Mizuno, and Rivaldo, the £30-million footballing star of Barcelona and Brazil. One expresses himself creatively with his head and keyboard, the other with his head and feet.

The project goes back to 1985, when Mizuno produced the Morelia, a boot which was significant in that it was a critical 50 grams (1.76 ounces) lighter than its contemporaries, and which was to become a mainstay at Mizuno for 14 years. Lightness and, to a degree, softness are crucial elements in a modern football-boot design, as this project highlights.

The foot is a complex set of 28 bones that are easily damaged and it is no surprise that protection and shock absorption are key factors. So how does one go about designing a boot which, when worn, makes you feel as though you are playing barefoot, yet protects the foot at the same time? The answer is through biomechanics and materials technology.

Lightness and softness in football boots are elements that are, arguably, best suited to the Brazilian style of play. The minimal but effective movements for kicking, changing direction, stopping and starting have become the hallmark of the Brazilian style, along with using only the lower part of the leg and foot to pass the ball. A lighter, more natural-feeling boot offers greater sensitivity, as well as a reduction of fatigue or injury and an increase in running speed and agility. The images of Brazilian children playing barefoot on the sands of the Copacabana beach one day, and wearing the gold shirt of Brazil in the World Cup (in heavily endorsed boots) the next, are familiar and evocative ones, but serve to emphasise this requirement of 'naturalness' in a contemporary boot. This holy grail of lightness, strength and softness is one that all manufacturers are seeking.

opposite bottom and side views of the Rivaldo Wave Cup football boot

MIZUNO

Scans of Rivaldo's feet were collated to produce a virtual 3D version, which then underwent simulated football play. The point of these tests was to find out which part of the foot is exposed to pressure, how the foot moves on impact and in what sequence the bones are articulated in action. This data, based on all patterns of play, were then deployed in the design development, which included the use of CAD (Computer Aided Design). Apart from a personal benefit to Rivaldo, a boot resulting from this type of research enables Mizuno to eliminate any unnecessary components of a shoe, while marshalling materials to the areas that matter most.

The results of these bio-mechanical studies were intriguing, as they revealed that the heel area, the area below the little and big toe, and the balls of the feet were particularly exposed. These are the areas in which pressure, over the course of a match, contributes most to a player's fatigue.

The response was to build in material and functional details that would alleviate these pressure points while retaining the boot's essential weight and softness. These included a lighter, more effective material for the laces (previously used in Olympic champion Carl Lewis's running shoes), an artificial leather mesh for the tongue, a honeycombed sole structure and nylon shoe base. The layout of the blades (studs) was developed for improved grip, the principal difference being the reduction to two main blades on the inside of the forefoot (there are usually four), the placing of blades further forward on the boot, and the appearance of two further blades right at the toe end.

This arrangement assists grip and traction, and helps prevent lateral movement of the pivot leg. The reduction in the total number of blades contributes to the overall weight saving.

As in running shoes, cushioning and stability are key requirements in a good football boot. This is achieved by incorporating mid-sole suspension, based on automotive thinking, which helps to avoid overpronation and the associated damage to ankle and knee joints, as well as minimising the collapse of the foot on landing. The suspension, called 'Compact Wave', maintains stability without increasing weight, and the honeycomb outsole has similar strength-to-weight ratio as a bird's bone. Handstitched kangaroo leather boot uppers continue the anthropomorphical references.

The weight of the first iteration of the new boot was about ten per cent less than that of the old one, 217 grams (7.64 ounces) per boot as opposed to 245 grams (8.63 ounces), but this wasn't enough for

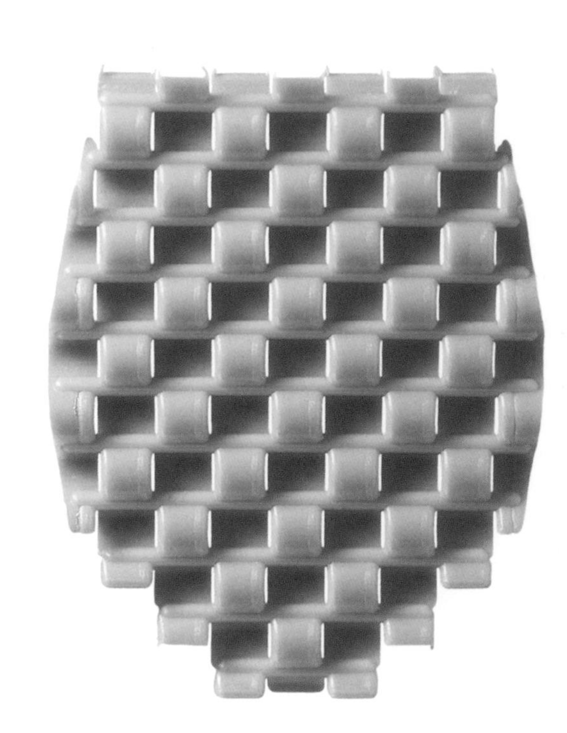

right Compact Wave Cup suspension insert
opposite top Rivaldo
opposite bottom biomechanical scans of Rivaldo's feet

either Mizuno or Rivaldo. A target of 200 grams (7.04 ounces) was set by Oorei, which was eventually achieved through further design and materials development. The specification now included an almost obsessive attention to detail, such as a thinner material for the Mizuno shoe stripe, a further change to the core of material for the heel and insole, a change in the lace material to 'Dyneema', a reduction in the amount of glue used and, incredibly, even a colour switch to a pearl white, to reduce the weight even further (as well as appearing lighter to the eye).

If any demonstration of the powerful combination of Mizuno and Rivaldo were required, it was at the 2002 FIFA World Cup in Japan and Korea. The newly crowned world champions, Brazil, featured a silver-shod Rivaldo, one of the major stars of that illustrious team.

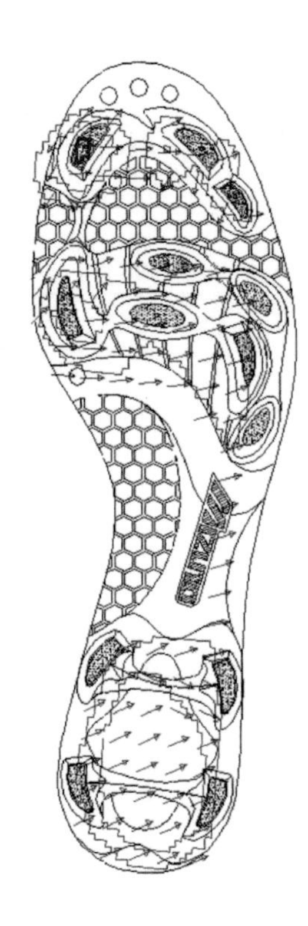

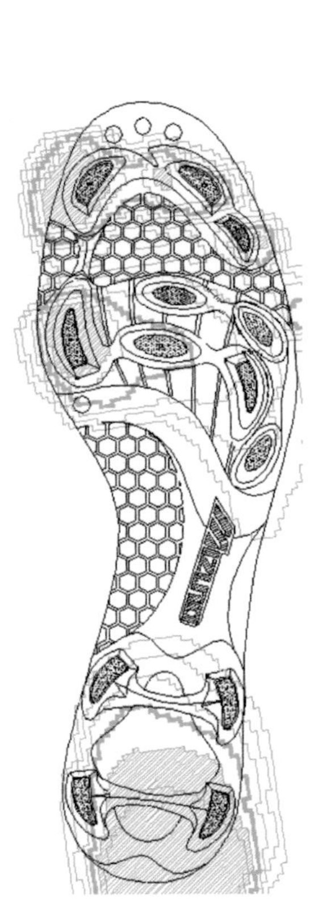

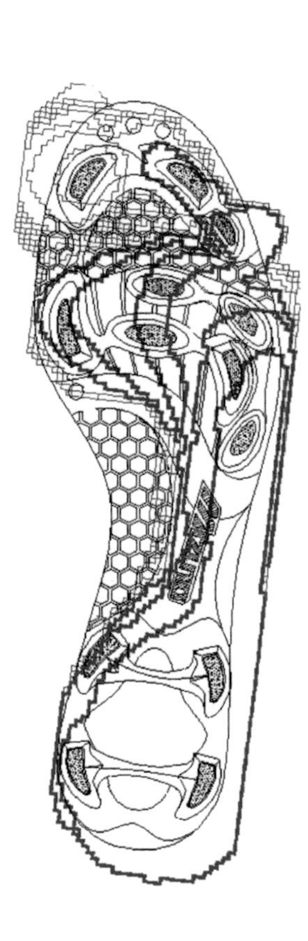

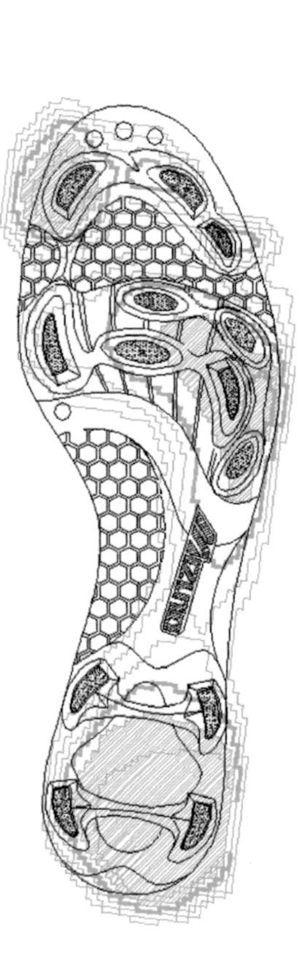

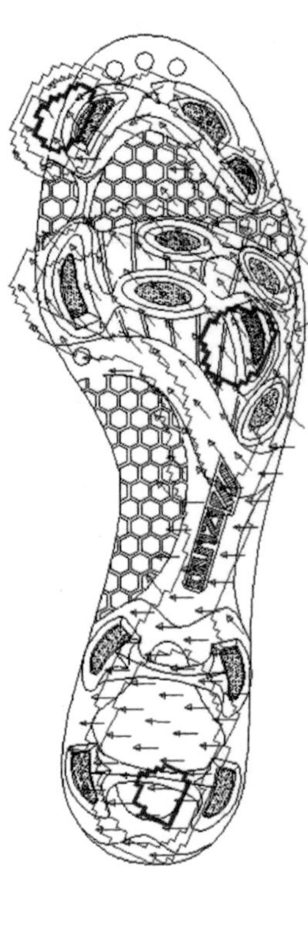

feint (right) heading instep instep crossing

Bridgestone Corporation

Potenza Formula One tyres

There are countless technological innovations from motorsport that now benefit everyday driving. The trickle-down effect of ideas, materials and applications from racetrack to driveway is nowhere more apparent than in safety features, like anti-lock brakes (ABS) and traction control. While better engineering and advanced technologies in motorsport are the obvious factors for these improvements, however, increasingly so, is chemistry.

The alchemy of modern Formula One tyre technology can result in the minuscule, but essential, difference between success and failure. The subtle mix of tyre compounds is calculated from race circuit to race circuit and specific weather conditions, from team to team. The sport is a complex political, economic and business juggernaut, where huge budgets approaching the size of the GDPs of some small countries are at the mercy of results. There is a close political and technical relationship between car manufacturer and tyre company which extends further and further into the development cycle of the car itself. Rather than being merely the 'boots' that are the physical contact between track and car, modern F1 car designers are building in the characteristics of each manufacturer's tyre to maximise the relationship between the dynamics of the chassis and the properties of the tyre itself. In effect, they are making the tyre an integral part of the car that is as important as the engine or gearbox. Bridgestone is at the forefront of the development of this matrix of car, track and tyre, providing the groundwork for dynamic improvements that will eventually lead to more efficient and safer road tyres and cars.

Each F1 dry-weather Formula One tyre, formerly called slicks, is required to feature four 2.5mm deep grooves around its 660-mm (26-inch) diameter. This important detail reduces the amount of grip between tyre and road surface, thus theroretically reducing the cornering speeds of the cars and increasing safety. When the car has completed the race there will have been a certain amount of wear on these tyres, but Fédération Internationale Automobiles (FIA) regulations state that these grooves must still be visible. The surface of a F1 tyre after it has completed the 60 or so laps of a race will be a tortured landscape of liquefied black rubber.

Tyres are designed to be optimised for the duration of a race only, sometimes just for a specific period of practice or qualifying, depending on which of the compound choices the team has made.

opposite Bridgestone Potenza Formula One dry tyre

POTENZA
BRIDGESTONE

Over the course of a season of 16 races, 40 dry-weather tyres and 28 wet-weather tyres are available to each driver over a race weekend. These are at their optimal performance at 90°C (190°F), and 50°C (120°F) for wet weather, hence the need for electric tyre-warmer blankets before the races start and during pit stops. Each F1 front tyre weighs approximately 9 kilograms (19.8 pounds) with each wheel rim weighing 3.5 kilograms; the tyres are 355mm (14 inches) wide at the front and 380mm (15 inches) at the rear, and run at between 17–22 psi.

The Bridgestone Corporation was created by Shojiro Ishibashi in 1931 (the name comes from a translation of the founder's surname meaning 'stone-bridge'), and since then has grown into a global player in the production of tyres for almost all applications, as well as bicycles and sporting goods like tennis racquets and golf clubs. The company's motorsport brands include Firestone, famous for racing in the USA, and Bridgestone Potenza in Formula One.

The Bridgestone tyre manufacturing process begins with the four main constituents of the tyre mix – natural rubber, sulphur, carbon black and a secret cocktail of chemicals. These elements, once combined, are extruded to form basic rubber sheets, which are cooled and reheated to increase the elasticity of the compound. This mix is extruded again and cut into sheets of the correct length (the tyre diameter). Spun cords of nylon, polyester and steel are then woven into the sheets and sealed in with rubber. Lengths of wire beading are also rubber-coated, formed into coils, incorporated into the woven rubber sheets and again rolled into tubes of the correct tyre diameter. The side tread rubber, steel belts and tread rubber are combined to form a 'green' tyre. After vulcanisation, the tyres are trimmed, balanced and finished, before being branded and sent to Bridgestone's Motorsport HQ in Langley, England.

Bridgestone began its first involvement with Formula One in 1997, at

above 2002 Formula One Constructors title won by Ferrari with tyre partner Bridgestone
opposite Bridgestone Potenza intermediate and wet Formula One tyres

the Australian Grand Prix in Melbourne, initially supplying tyres to Lola and subsequently to five more F1 teams that season, resulting in a fifth-place finish. Since then it has won four World Constructors Championships, was sole supplier of tyres to F1 in 1999 and 2000, and achieved its 50th Grand Prix win in 2001. The Japanese GP at Suzuka in October 2002 marked Bridgestone's 100th F1 Grand Prix.

Bridgestone currently supply their race products to Ferrari, Arrows, Jordan-Honda, Sauber-Petronas and BAR-Honda. Four huge Bridgestone transporter trucks are required to carry up to 1600 tyres to each race to meet the combined needs of these teams. The Scuderia Marlboro Ferrari team is currently Bridgestone's chief development partner, and the F2003 cars are developed with a strategic exchange of data between car manufacturer and tyre manufacturer. This relationship produces a highly competitive package where tyre dynamics are as critical as aerodynamics. More than ever before, the tyres on an F1 car are as critical a part of the car's behavioural characteristics as the engine, chassis, gearbox or driver – a fact amply demonstrated by Bridgestone and Ferrari's domination of the 2002 season.

Shinkansen

High-speed trains

Japan is the birthplace of the high-speed railway, and the Shinkansen has provided over 35 years of fast, safe, reliable service. Japan's pre-eminence in this highly technical field is amply illustrated by the current Shinkansen, as a benchmark for train systems the world over, and by on-going research into next-generation Magnetic Levitation trains. But the Shinkansen is more than just the fastest, most efficient and safest train system ever conceived: it occupies a central place in the psyche of the Japanese people – as potent an icon in Japan as Concorde is in Britain or France, or the Apollo space programme in the USA.

Originally launched in time for the Tokyo Olympics in October 1964, the Shinkansen (meaning new main line) linked the main industrial cities of Tokyo, Nagoya, Kobe and Osaka, travelling at up to 200kph (125mph), a phenomenal speed for a regular train service in the 1960s. The construction of the new railway was a huge undertaking, part of a larger transport infrastructure development in preparation for the Olympics. However, the roots of the concept went back almost 30 years.

Before the Second World War a similar plan for a high-speed rail system had been initiated but never completed, to connect the industrial centre of Japan with China, Korea, Russia and ultimately Europe. Engineering problems, the geography of the country and eventually the outbreak of war led to the plan being abandoned. During the late 1950s, a version of the project, led by Dr Shima, the 'father of the modern Shinkansen', was revived as a way of joining the four main islands of the Japanese archipelago. This new, high-speed rail link was, initially, planned to carry both passengers and freight, but as the popularity of the system increased, freight was dropped in favour of an all-passenger service.

The 1964 Tokaido Super Bullet Train represented the beginning of the end of the rehabilitation of Japan, both emotionally and economically, bringing with it a huge surge of national pride. The Shinkansen was a marvel of engineering. Economically, it allowed the newly-burgeoning business culture to connect speedily between major industrial centres within hours. In 1958, a journey from Tokyo to Osaka would have taken 6.5 hours; in 1965 it took 3.5 hours.

opposite Series 700 Shinkansen speeds past Mount Fuji

It was a huge engineering feat to produce a fast, level surface for the high-speed tracks, as 21 per cent of land in Japan is uninhabitable mountains and hills. In some cases earth around the tracks was found to be relatively soft, which meant rethinking the layout of the heaviest parts of a train, the engines. The existing convention was to place an engine at either end of the carriages that made up the train, but a way was found to redistribute the weight of the power units into smaller engines in all the carriages (called a split power unit, or SPU, a system now used by German railways). The engineering achievement represented by the Shinkansen provided a valuable sense of self-belief for the Japanese public in their ability to create world-class technologies.

Series 700 Shinkansen

Train spotting, or standing on train platforms recording railway data, is a well-established pastime all over the world, but this kind of enthusiasm, especially for accurate linear movement, is more comprehensible in Japan than elsewhere. The Shinkansen represents the epitome of modern rail excitement, providing visual drama and punctuality. Standing at the far end of a Shinkansen platform in Tokyo station, it is possible to experience the beautiful and harmonious development of Shinkansen trains from the early 100/200/300 series (still running), through to the latest 500 and 700 series.

All Shinkansen are aluminium-bodied for the weight-saving that is vital for efficiency and speed. Aluminium is also a key material for the many handmade parts on bullet trains which require special skills in assembly and manufacture. For example, manual construction techniques are required to build the organic, aerodynamic front-end of the 700 and only a handful of skilled craftsmen remain in Japan able to manipulate these svelte aluminium forms.

The Shinkansen is an extremely popular choice for travel in Japan, with a top speed of 270kph (168mph), a door-to-door time-saving over air travel between major cities. Since its inception in 1964, over six billion passengers have travelled by Shinkansen without a single reported incident. This popularity puts a strain on infra-structure, however, with almost 65 per cent of all costs sucked into track and train maintenance. Although most Shinkansen run on conventional steel tracks on concrete sleepers (1067–1435mm gauge), some have bespoke lines to avoid possible conflict with more conventional, slower trains. There are three types of Shinkansen: the swiftest, Nozomi (Desire), Hikari (Speed of Light) and Kodama (Echo). Nozomi are recognisable by the long streamlined snouts of the 500 and 700 series trains with aero-dynamics that offer slippery air-coeffciency, minimal wind noise (inside the carriages and in the surrounding area) and an unmistakable profile.

The design of the original 1964 100 series closely resembled an airliner-without-wings, a pure, streamlined projectile; but the compelling aerodynamics of the current 700 series stirs real excitement and anticipation of a special journey. The nose of this sleek, purposeful beauty can be glimpsed edging its way out of the claustrophobic landscape of downtown Tokyo. Designed in 1999, it is perhaps the most startlingly dynamic train ever created, at it's most dramatic when sweeping into a station, flashing past rice fields, or most famously, beneath Mount Fuji. The engine drivers congregate on the station platforms with all the insouciance

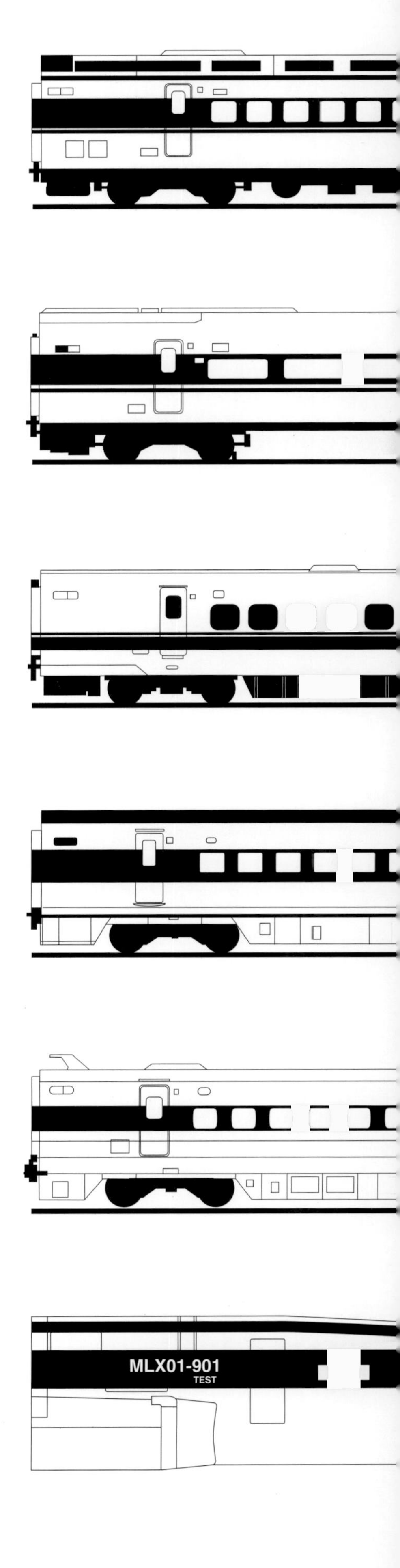

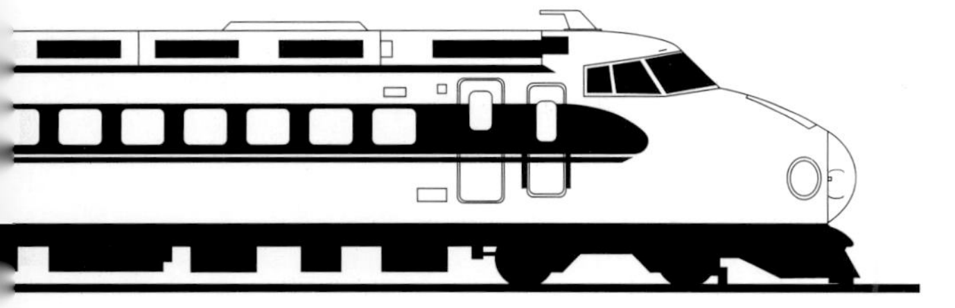

JR Series-0 'Nozomi' Shinkansen from 1964

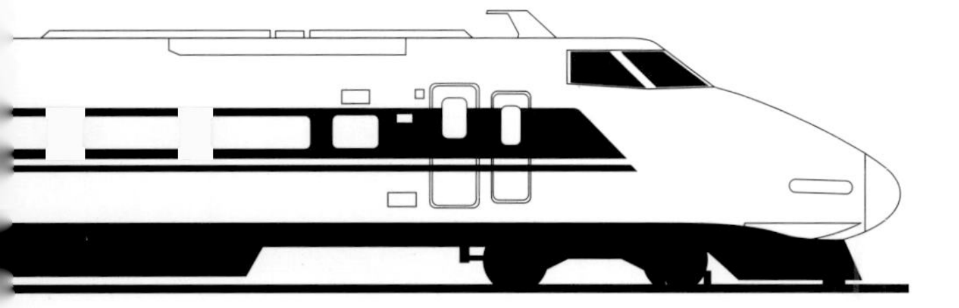

JR Series 100 'Hikari' Shinkansen from 1985

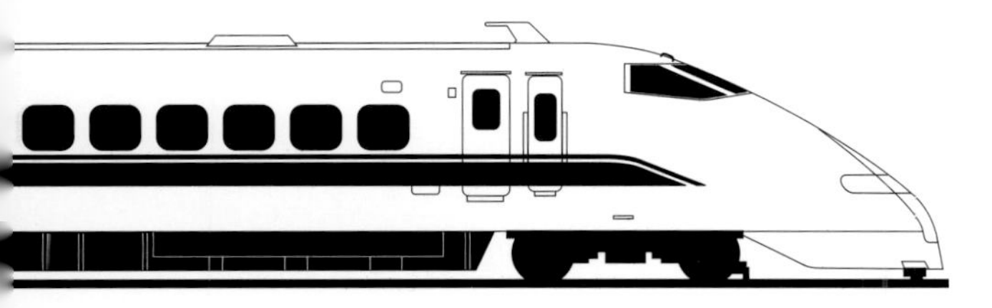

JR Central Series 300 'Nozomi' Shinkansen from 1992

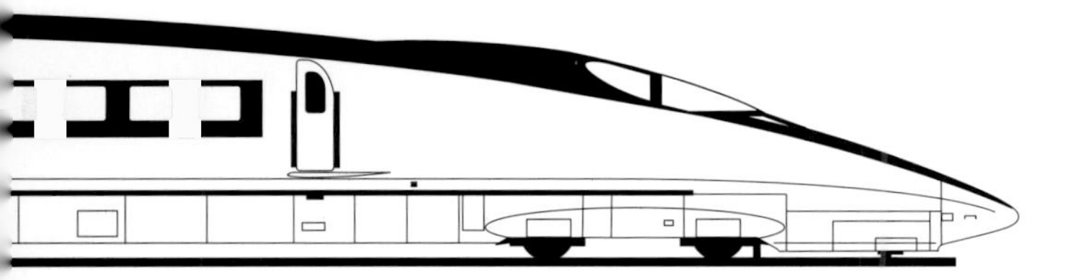

JR West Series 500 'Nozomi' Shinkansen from 1992

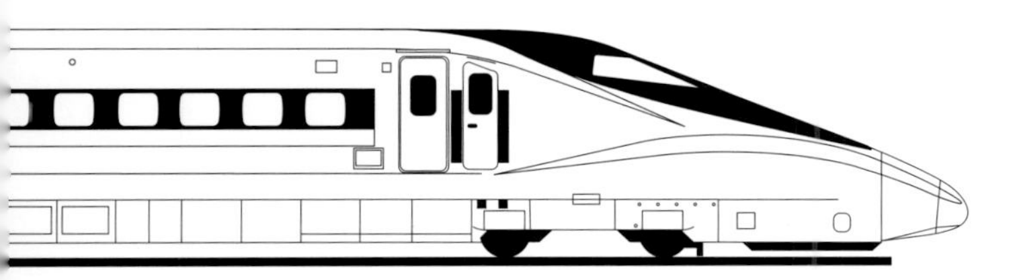

JR Central/JR West Series 700 'Nozomi' Shinkansen from 1999

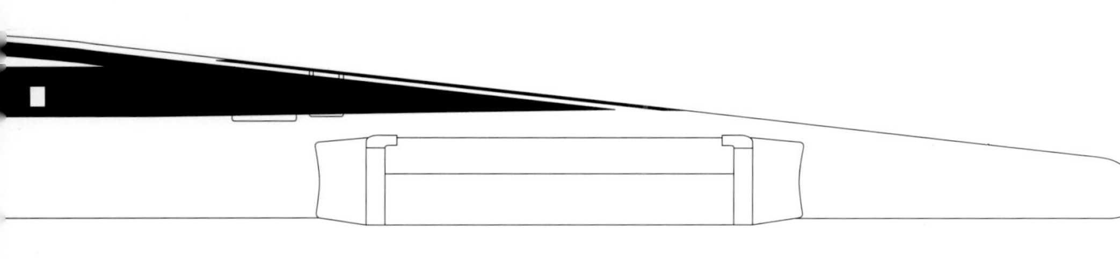

MLX01-901 maglev linear experimental train from 2002

and swagger of airline pilots, justifiably filled with pride as their beautiful chargers arrive, gleaming and white, ready for another ride into the future.

The series 700 Shinkansen is perhaps one of the final iterations of the Shinkansen before the Maglev makes its debut in around ten years' time. Research into Maglev rail travel began in Japan in 1962, two years before the introduction of the conventional Shinkansen, and has since cost approximately $2.5 (£1.8) billion of public money - a significant investment on behalf of the Japanese people by the government and train companies. In April 1994, a new world rail speed record was set at 552kph (368mph) by a manned research Maglev train in Japan, which suggested that a journey between Tokyo and Osaka (currently taking over two hours), might be completed in just one hour, when the service is eventually introduced sometime this decade or early next.

In so-called Maglev (meaning 'magnetic levitation') rail travel, a huge magnetic field is created between the train and track, and the repelling and attracting forces of on-board, superconductive, niobium-titanium alloy magnets, which are cooled by liquid helium, are used to propel the train forwards at super-high speeds. Apart from retractable rubber wheels that are used just to start the train rolling, no part of the train touches the track. As such it creates no friction, very little noise, has a correspondingly low environmental impact, and gives travellers a very smooth, fast ride. The extraordinary aerodynamic shapes that have been seen on 'conventional' Shinkansen trains have lead to some beautiful and inspiring shapes: the current test-train, the MLX01-901, is a study in aerodynamic beauty and sharp intent.

OLYMPUS
O·product
OLYMPUS LENS 35mm 1:3.5

3 the beauty and the best

THE AESTHETICS OF NEW TECHNOLOGY

Katachi is a traditional Japanese concept expressing 'the marriage of beauty and functionality', an approach that is perfectly illustrated in the contemporary context by Yamaha's award-winning Silent Violin and Cello. These new forms of traditional instruments allow players to practise their music without causing disturbance, important in densely-populated environments, whilst gaining the confidence and expertise to perform. This range of products captures the harmony of sound and form, juxtaposes the qualities of old and new and expresses the beauty of sensitive engineering and design.

Canon's EOS-1D digital camera is a great example of flawless functionality twinned with a perfect ergonomic form which again celebrates the notion of *Katachi* in a modern, highly technical, leading-edge product.

Possibly the best racing motorcycle ever conceived, the Honda RCV211V, is a world-beating racing machine which has almost literally disappeared off the track, pushing the boundaries of technical excellence so far that its rivals are stunned by its near perfection. The single objective of the RCV211V is to be ridden to victory. Its function is focused and its raw beauty has been derived from that very function.

All these examples, of course, require an essential human involvement. The instruments, the camera and the motorcycle are created to enhance, in some cases even exaggerate, the skills of the player, the photographer or the rider. These products are not simply inanimate objects, they are highly functional tools, requiring the skills of the user to bring them nearer to perfect performance and in doing so, celebrate the meeting of man and machine.

opposite Olympus 'O' Product 35mm camera

Canon Inc.

EOS-1D digital SLR camera

Digital photography, or more accurately, digital image capture, has long been seen as an example of the double-edged sword of technology, rather like the vinyl-versus-CD debate of a few years ago. On one hand, the digital capture of images means that there has never been a more exciting time for collecting and sharing images easily, quickly and with real spontaneity. On the other hand, *aficionados* of the traditional film process feel that the subtle skills of exposure, composition and execution are central to photography and are impoverished or even lost in digital image capture. Both points of view are valid, but as digital camera technology advances and comes closer to producing the qualities of the traditional process, the cherished benefits of conventional photography begin to recede.

The professional digital SLR camera is an example of this continuous technological assault on traditional techniques. In professional photography, there is very little space for emotional attachments to tradition: as soon as digital cameras offer better usability than their technological forebears, they will be adopted. That point has now been reached with the EOS-1D.

The 2001 flagship of Canon's professional digital range, the EOS-1D is a thoroughbred SLR camera, and is the epitome of current professional digital cameras in a flexible, accurate and usable package. The ergonomically sculpted body is water- and dust-resistant (the 72 major body seams and all moving controls are gasketed and sealed), rigid and durable. It is an important factor when designing to meet the demands of professional sports photographers and photojournalists. Internally, the magnesium-alloy structure and aluminium mirror box are light and strong, and carbon-fibre shutter blades are used for greater durability and reliability – the EOS-1D shutter can operate up to 150,000 cycles. While a compact body may not seem as crucial a feature in a professional camera as in a consumer product, smaller dimensions are in fact a constant request from professional photographers who often carry two or three camera bodies at once. Small size and light weight are key points of the EOS-1D design.

opposite Canon EOS-1D digital SLR camera

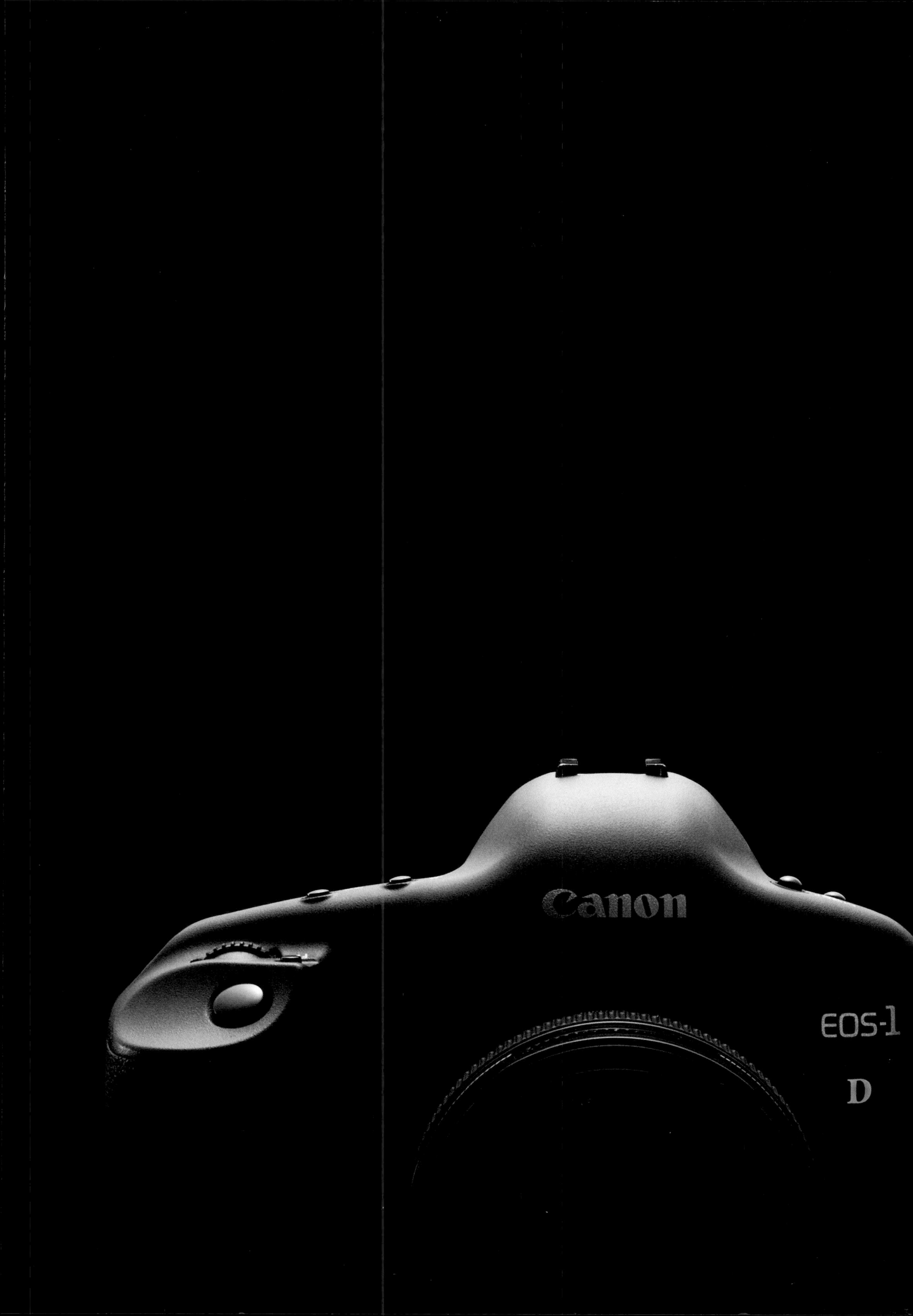

Canon

EOS-1

D

The control layout, user interface and ergonomics are designed to be the same in both the digital EOS-1D and the analogue EOS-1V camera bodies, so that photographers who carry both formats can switch easily from one to another. Attention to detail also means that Canon has made controls and dials easy to find, but not too easy to adjust, to avoid accidental or mistaken setting changes, another user-driven requirement.

In this competitive profession, the instant capture and transmission of a high-quality image (up to 4.15 megapixels on the EOS-1D), to sites worldwide is of the utmost importance and is now central to contemporary news image gathering. The EOS-1D, the world's quickest digital camera when launched, can shoot 21 consecutive frames at up to 8 frames per second. The images can be stored on compact flash cards (CF cards) or on micro drives and laptops. There is even a built-in microphone for photographers to add audio annotation to their images. The camera is Firewire-enabled for data transfer at up to 400 mbps (megabits per second). Images can be adapted or adjusted before storage or transmission through image editing software and saved in TIFF, JPEG or RAW image format. Of course, Canon's range of lenses, scanners, printers and peripherals are all compatible, making the EOS-1D less of a camera and more a high-performance professional digital imaging tool.

right magnesium alloy EOS-1D camera bodies
below EOS-1D is water and dust resistant
opposite Canon EOS-1D

Canon
CANON LENS EF 50mm 1:1.4
CANON LENS MADE IN JAPAN
EOS-1
D
DIGITAL

Yamaha Corporation

Silent Instruments

In Japan the qualities of harmony, simplicity and calmness are encapsulated in Zen. Yamaha's approach to product development is guided by this philosophy. The company also believes in 'innovention': in this case, redeveloping a traditional musical instrument into something new and unique while retaining the essential nature of the original.

Sympathy with one's surroundings is the starting point for Yamaha's Silent Instruments. To play any instrument, even to the most basic level, requires practice – and understanding neighbours. In a country like Japan, it is impossible to practise any instrument without letting others know your skills or limitations intimately. Taking this reality of urban life as a starting point for developing a socially responsible silent stringed instrument, Yamaha also took the opportunity to reinvigorate the violin, cello and bass, evolving them into more flexible, approachable digital instruments for beginners and experts alike.

The results arguably represent the first significant shift in musical-instrument design for literally hundreds of years. By the judicious use of modern technology, Yamaha's engineers have been able to reproduce all the essential aural qualities of a traditional stringed instrument; at the same time these are objects of real beauty, even to non-players.

opposite Yamaha SV-100 Silent Violin

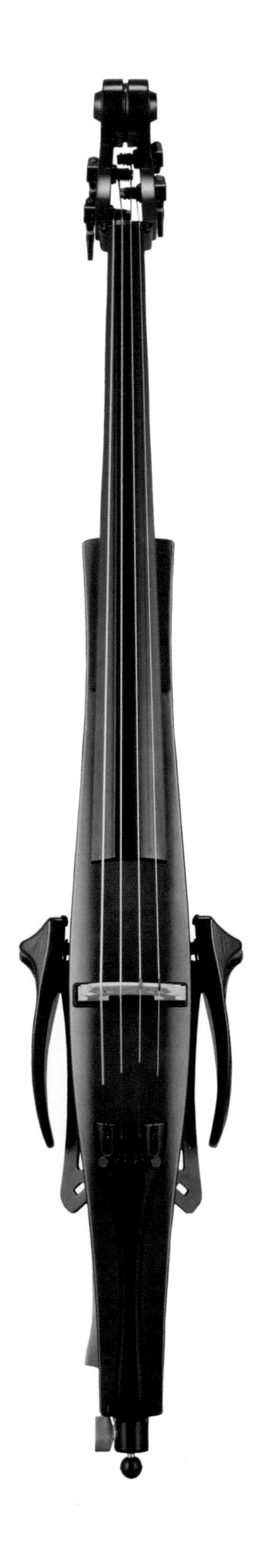

SV-100 Silent Violin

On an acoustic level, to produce a clear sound on the violin one has to appreciate the correct musical intervals. On the silent instruments, clear sounds are generated as a result of the use of particular materials: ebony, maple and cedar wood for the finger board, bridge/neck and body respectively, and polymers for the side body, chin rest and tail piece. The rich reverb is enhanced by an electronic chip, which can be channelled to simulate large-hall, medium-hall or room-size environments. This gives players the ability to control both instrument and environment, and allows them the confidence to practise without hindrance and to play in concert-hall conditions at will. Equally, by connecting it to other digital equipment, one can play along with specialist training or other favourite recordings. As well as with a warm, full-bodied tone, the Silent Violin possesses most of the playing characteristics and nuances of any traditional violin.

Just as compelling is the minimalism of the bodywork. Because tone and reverb are produced digitally, there is no need for a resonance chamber. But Yamaha realised that to dispense completely with the signature form of the violin would be a mistake both functionally and visually. The traditional shape of the right side of the body has been retained to enable the player to achieve a correct bow angle and to ensure that the instrument keeps its essential visual identity.

SVC-200 Yamaha Silent Cello

The traditional cello is a sensual instrument with its own unique tonal spectrum, and one that demands physical contact. The Silent Electric Cello has slimmed down the bulky form of the traditional instrument, preserving its spatial and acoustic qualities while offering a surprisingly compact design. The vibrations of its strings are traced by sensors and converted into digital signals that allow the player to hear and experience the full warmth and timbre of the acoustic instrument while producing only one-tenth of its volume. Like its smaller cousin, the Silent Violin, the Silent Cello has lost its traditional resonance chamber and the digital signals produced can be controlled and channelled into a choice of environments. This gives it a unique pared-down visual appeal, and greater simplicity in that the residue of the body can be folded up to make a truly transportable instrument.

opposite (left to right) the Yamaha Silent range also includes the SVG-100 Silent Guitar and the SV-100 Silent Violin
this page the SVC-200 Silent Cello

Honda Motor Co., Ltd

Honda RCV211V MotoGP racing motorcycle

Sports Wing – Xasis concept

Of all the Japanese motorcycle manufacturers, it is the name of Honda that resonates the loudest around the racetracks and roads of the world. A strong corporate culture, a solid foundation of technological and engineering excellence and an unprecedented number of race wins have combined to give the company a profile that is second to none. Honda's technical virtuosity feeds directly from the workshops of Hamamatsu on to the racetracks, and from there on to the high street.

For Honda, the motorcycle revolution began with the Model A in 1947. Conceived by company founder Soichiro Honda, this was essentially a motor-driven bicycle. But the Honda racing success story starts in 1961 with the 125cc RC143, ridden by Tom Phillips, which clinched Honda's inaugural Grand Prix win in Spain. This high-revving, four-stroke DOHC parallel twin beauty even had its own signature sound, a crisp but powerful wail which became known as 'the Honda music'. The first Japanese motorcycle company to take part in the motorcycle racing world series, Honda quickly established a winning formula that it has maintained to the present day. In 2001, it won its 500th world Grand Prix at Suzuka in Japan with the NSR500, a two-stroke, liquid-cooled V-4 500cc works machine, ridden by Valentino Rossi. Between the RC143 and NSR500 landmarks, Honda has produced a series of bikes that have constantly pushed at the edges of technology and engineering, and sometimes beyond. The names of those who have sampled and harnessed the potential of these bikes make up a roll-call of the best riders in the world, from Mike Hailwood in the 1960s and Freddie Spencer in the 1970s, to Valentino Rossi in the present day.

opposite Honda RC143 racing motorcycle from 1961

60
HONDA
60

It is the enigmatic Rossi who will be taking Honda on to the next stage of its racing successes in the revolutionary MotoGP world series - the new premier racing class - with the RCV211V. This is a truly next-generation racing tool powered by the unconventional 340kph (204mph), 220bhp, 990cc liquid-cooled, four-stroke, four-valve DOHC (direct overhead camshaft) 75.5-degree V-5 engine, designed by Tomoo Shiozaki, project leader at HRC (Honda Racing Corporation). A five-cylinder engine seems an unusual configuration for a motorcycle. Although five-cylinder car engines are familiar enough, they have rarely crossed over to motorbikes. However, Honda has been experimenting with this five-cylinder layout since 1965 with the 125cc RC148, a high-revving, lightweight in-line 32bhp, four valves-per-cylinder engined bike capable of around 208kph (125mph). A version of this bike, the RC149 ridden by Luigi Taveri, was the 1966 125cc World Championship-winning bike.

The RCV211V is a more than worthy distant relative of that sensational bike of the 1960s. As well as the V-5 engine, all major aspects of the bike - including the chassis and aerodynamics - were developed simultaneously, starting with a clean sheet of paper, to produce an integrated and complete design; a great example of creative and original engineering thinking. This approach is demonstrated by the integrated aerodynamics, represented by the distinctive low frontal area.
The designers have aimed at overall aero-dynamic efficiency rather than merely reducing air drag. A racing motorcycle differs from a racing car in that a low-drag resistance, high-downforce bike will be efficient on the straights but becomes difficult to handle around corners.

right RCV211V V-5 engine
far right Honda RCV211V front and back
opposite Honda RC143 with RCV211V in foreground

With a rider of Rossi's diminutive size, it has been possible to reduce the size of the frontal area as well as the overall size of the bike, providing an excellent balance of straight-line speed and cornering stability.

The RCV211V is possibly the greatest racing motorcycle yet, and continues the glorious legacy of its predecessors by winning the inaugural MotoGP series world championship in its first year of competition.

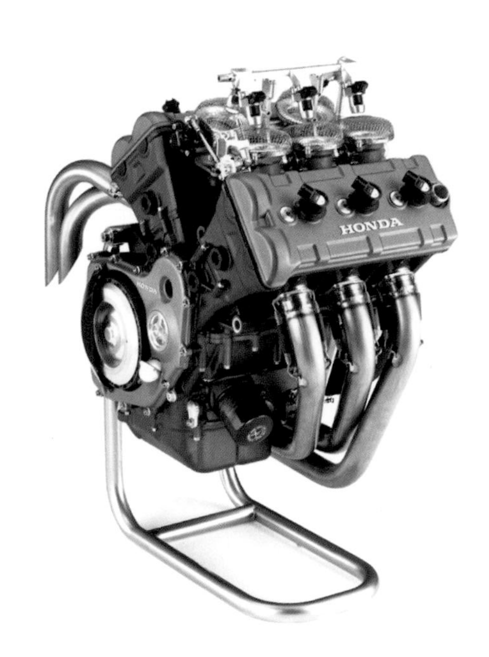

60
HONDA
60
HONDA
HRC
HONDA
MICHELIN

above 2002 MotoGP world champion Valentino Rossi
right top Rossi and teamate Tohru Ukawa on RCV211V bikes
right bottom Valentino Rossi racing at Cataluna in Spain, 2002

Sports Wing – Xasis concept

This handsome Xasis prototype is an indication of the direction Honda could be taking track and road bikes – towards a brutal functionalism in which power and technology are put firmly and defiantly on display. This future-focused bike has its radiator built into the tail fin, an aerodynamic undercowl silencer and a fairing with styling straight out of Gundam science fiction. Mechanically, a 1,000cc, four-stroke, 90-degree V-twin is shoehorned into an aluminium twin-spar frame, and single-sided front and rear suspension adds to the visual impact. A huge rim-mounted disc brake on the front matches the engine power with huge stopping power, and the hunched, compact and uncompromising styling carries all the passion and progression that the marque represents, blurring the boundary where racetrack meets road.

this page Honda Xasis concept motorcycle study

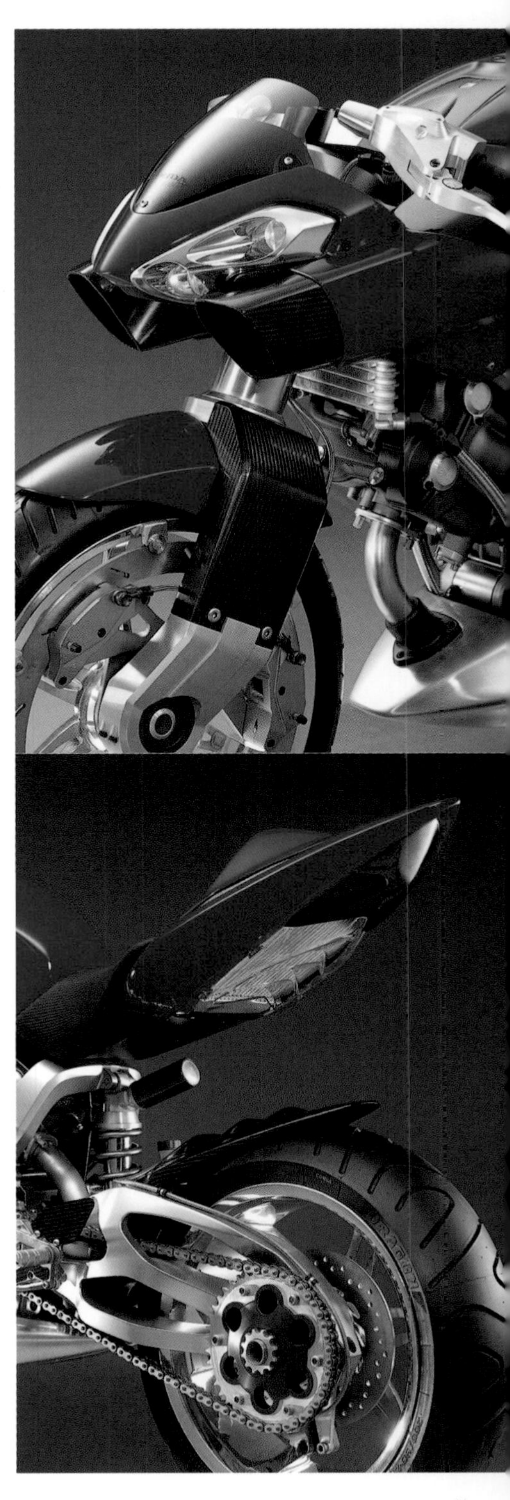

Sharp Corporation

Aquos display monitors/televisions

For a company centred on a particular technology, Sharp has an unusual design strategy. The technology is LCD (liquid crystal display), and Sharp's unusual design strategy is that it does not really have one, or at least not in a conventional or recognisable corporate form. The company has a long and illustrious history of product development: from the first LCD calculator in 1973 to the current Viewcam camcorders, it has continually created products that could be described as state-of-the-art technology with a discerning face, that is, where the product design is not intrusive or overpowering. The personality or character of an electronic product is perceived by a consumer through its interface (software), conveyed by its display, and it is in this crucial area that Sharp's software engineers have spent time perfecting technologies such as handwriting recognition. This signature interface is the key to retaining a loyal consumer base, and of course communicates the company's design identity and values.

The Aquos monitors are a successful vehicle for the LCD technology in which Sharp leads the world. Designed by Toshiyuki Kita and Sharp's in-house design and engineering team, this range is distinctive and stylish. The Aquos monitors are a welcome departure from the plain office-orientated LCD monitors of the past; the soft, organic feel of the facia and the foot styled with anthropomorphic élan represent a fresh direction in this field. This kind of innovative approach will become much more evident in the

opposite Sharp Aquos display monitor

AQUOS
SHARP

小 音量 大
∨ 選局 ∧
チャンネル設定
入力切換
電源
入
切
AQUOS

future as Japanese companies try to combat cheaper 'me-too' products from other parts of Asia with distinctive and original design.

Interesting also are the concessions to the problem of product disposal. The non-halogen cabinet material produces no dioxins when incinerated; to aid recycling no chrome plating has been used internally, and recyclable polystyrene has been used in the stand.

The 28-inch LC-28HD1 LCD monitor/television is the flagship of the range, and is a low-reflection, high-resolution screen with a wide (170-degree) viewing angle. The challenge of creating something exciting in this competitive field was met by this strong, confident and influential design, which has proved to be an overwhelming commercial success.

opposite Sharp LC-28HD1 LCD widescreen television
this page Aquos display monitors

Canon
WP-DC300
WATERPROOF CASE
ZOOM LENS
7.1-21.3mm
30M/100ft.

4

miniaturism

MINI PRODUCTS, MAXI FUNCTION

The miniaturisation of form twinned with the maximisation of function is a Japanese speciality, developed over decades. It is a hallmark of Japanese design - one of the main parameters for a Good Design award in the 1960s was that an entry should demonstrate size reduction as well as performance. Today there are many products of gorgeous jewel-like proportions offering huge levels of performance.

This chapter will explore the complexity and engineered beauty of the miniaturised products that inhabit our everyday lives, products that we frequently take for granted. They have a precision and technical virtuosity that usually go unseen and are sometimes unappreciated. Examples such as Canon's IXY cameras are demonstrations of the power of complex technology in such a convenient, intelligent and frankly beautiful, package, capable of playing an important role in our lives and memories.

There are hundreds of other examples, but here we consider just a few of the best. Sharp's Muramasa laptop PC takes its inspiration from traditional Japanese swords, but the miniaturisation of its components and precision-engineering is firmly 21st century. Seiko Epson's EMRoS robots, intelligent computerised automatons, are just 1cc in volume but contain 120 separate components miraculously built by hand. Casio's G-Shock watches, with incredible functionality crammed into a popular, wearable product, perhaps represent the epitome of the merger between professional and consumer technology. For a country that has embraced mobile communication like no other, NEC's i-mode phones are ground-breaking personal products, pioneering the successful clamshell format in a beautifully crafted and tiny package.

opposite waterproof case for Canon Powershot S40 camera

Canon Inc.

IXY APS camera

IXY 200a/300a digital cameras

When compact 35mm cameras changed from metal to predominantly plastic bodies in the 1980s, it seemed the market was heading towards the dead-end of price-driven features and superficial styling that blights so many consumer-product markets. With the IXY camera, Canon helped to wrench the industry back to tactile, lightweight metals, producing a truly epoch-making product.

Other manufacturers had used exotic materials such as magnesium and titanium for camera bodies as early as the mid-1980s, usually for professional cameras, but it was Canon that introduced the first truly significant, consumer-orientated metal camera, the IXY (IX240 APS) in 1996. With its superb detailing and chic style it shook the foundations of the industry. Along with the perfect timing of its launch – just before the arrival of digital cameras – the IXY pressed all the right emotional buttons at the right time and became the unchallenged camera success of the mid-to-late 1990s.

opposite Canon IXY 300a and 200a digital cameras, 2002
top two shown at actual size

Canon
CANON ZOOM LENS 3x 5.4-16.2mm 1:2.7-4.7
2.0 MEGA PIXELS
IXY
DIGITAL

Canon
IXY
DIGITAL
LENS 5.4-10.8mm 1:2.8-4.0
2.0 MEGA PIXELS

Canon

A seminal product in Canon's history, the original APS (Advanced Photo System) format IXY was the beginning of the reinvention of the compact camera market and the redefinition of the camera as a carry-anywhere, image-defining accessory; as familiar and essential as a piece of jewellery, watch or lighter. The charming, covetable IXY appeared around the necks of celebrities from Tokyo to London to LA and on the pages of glossy magazines everywhere, and quickly became a style icon of the 90s. So great was its impact that it seemed compact cameras would never be the same again. After the launch of IXY, cameras became objects of desire, products of engineering excellence, and as a result, the market grew exponentially.

The consumer-friendly features built into APS cameras meant new opportunities for designers as well as users. The relatively small new film casings provided greater freedom in the design of more compact and sophisticated bodies. And for consumers, the original IXY featured drop-in film loading, a three print-type control (CHP), a 24-48mm zoom lens, retractable flash and back-of-print title imprinting. But it was the diminutive, tactile beauty of its stainless-steel body that inspired such passion for the camera. It was the first camera to influence its market sector by setting new benchmarks for finish, material and desirability, and went on to set an important stylistic trend for many other products, such as mobile phones and personal audio products.

right original Canon IXY APS camera, 1996
opposite IXY 300 digital camera
overleaf IXY 200 digital camera

Canon
2.1 MEGA PIXELS
CANON ZOOM LENS 3x
5.4-16.2mm 1:2.7-4.7
IXY DIGITAL
300
SET
WB
MENU
DISP
Canon
CF OPEN

CANON ZOOM LENS 5.4-10.8mm 1:2.8-4.0
Canon

CANON ZOOM LENS
IXY
DIGITAL
AiAF

Sharp Corporation

Mebius 'Muramasa' notebook personal computer PC-MT1-H3

In 1979, during the so-called 'electronic calculator war' in Japan, companies were competing to build and produce the world's thinnest calculator. Sharp succeeded with the 1.6-mm (0.06-inch) EL-8152, an early celebration of the new values promised by the microchip, and still an impressive feat today. A modern example of this pursuit of thinness is the Muramasa Notebook PC from Sharp.

The Sharp Mebius PC-MT1 laptop PC is the world's lightest, at 1.31kg (2.86 pounds) including batteries, and thinnest (16.6mm or 0.60 inch). It represents the ultimate expression of progress in LCD technology. The development of the slimmest mobile notebook PC has been an on-going quest for the holy grail of simulating an electronic sheet of paper. This chic laptop is at the cutting edge of design and technology in more ways than one. Named Muramasa after a legendary Japanese sword-maker famous for the sharpness of his blades, this product is a statement in Japanese engineering virtuosity that represents all the values that Sharp is famous for – LCD technology, usability and innovation.

There are superb details in this anorexic addition to the executive armoury, including a keyboard which is raised and lowered as the laptop opens and closes, and a low-reflection XGA TFT LCD display. In the pursuit of both thinness and toughness, the chassis is crafted from a single-piece magnesium 'box-rahmen' (rahmen means 'frame' in German) construction for light weight and strength. The magnesium lid and casing are dyed rather than painted silver, for durability, with other crucial architecture and components made in aluminium-alloy. The LCD screen is bonded directly on to the lid enclosure, eliminating the conventional framework needed to support the LCD.

Apart from its powerful specification, the laptop has a battery that stores up to three hours of power, or nine hours with an external power pack. All drives (for DVD/CD, smart cards, PCMCIA, etc.) are external – the price for such a svelte product.

opposite Sharp Muramasa notebook computer

Sharp is interesting from a corporate point of view for two reasons: the usability of its products, derived from a concentration on ergonomics and software interface design; and its refusal to adopt a single product design identity, concentrating on innovation instead. Sharp introduced the world's first transistorised calculator in 1963, and in 1973 the first LCD calculator (the EL-805; see below). Since then it has been the leading developer of LCD technology. Sharp is also the kind of Japanese company for which the maxim, 'the only technological constant is that it is always changing' is central. As such, it will always have the edge when it comes to innovation. The technologies involved in this laptop move quickly: by the time it is widely available, no doubt the title 'world's thinnest mobile PC' will have been claimed by an even thinner, lighter and more powerful version.

The Sharp EL-805

Launched in 1973, the EL-805 pocket calculator was the first product to use the new LCD technology, which has become the mainstay of Sharp's business over the past 30 years.

The late 1960s had been a period of considerable economic tension in Japan, and many manufacturers were cautious, but Sharp chose to innovate by developing a totally new product. In its diminutive way, the EL-805 calculator represented more than just a short-cut to accurate calculation in your pocket. In retrospect, it represents a more innocent time, when technology was less complex and when 'convergence' meant nothing more than putting a radio and a cassette player in the same box. Industrial designers were beginning to mark out the territory for specific new products, creating a character or personality for these gadgets for the first time, the lineage of which can still be seen in many of their modern descendants today.

Designed by Matafumi Ikeda, the EL-805 calculator was a revolutionary product, an aluminium masterpiece of industrial design that was $^{1}/_{12}$ the thickness of its contemporaries, $^{1}/_{25}$ the weight, used $^{1}/_{250}$ the number of parts, and used only $^{1}/_{100}$ of the power. Predictably, the price of this leading-edge product was huge - ¥26,800 (at current exchange rates, about US$400/£300) - but it automatically assumed a status-symbol presence on the desk of many executives, and to a degree, embodied the same kind of qualities and aspirations that laptop computers, mobile phones and PDAs do now. In fact, the world's first electronic organiser, the PA-7000 of 1987, was a direct descendant of the EL-805.

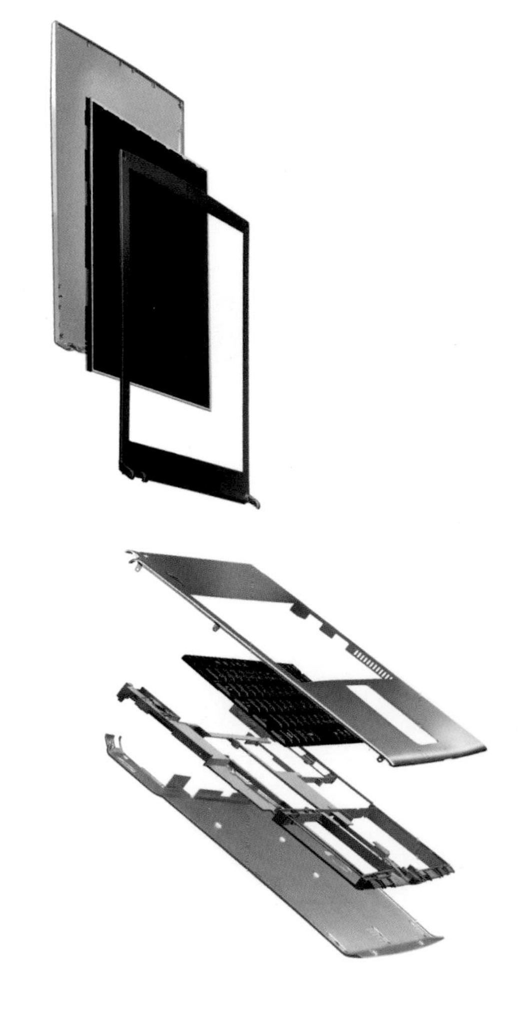

top right exploded view of Muramasa's box-rahmen construction
right EL-805 (1973) and EL-8152 (1979) calculators (left to right)
opposite the Muramasa laptop is only 16mm thick

SHARP

Seiko Epson Corporation

EMRoS (Epson Micro RObot System)

Seiko Epson is renowned for making a wide range of world-class information products: advanced ink-jet printers, electronic devices, semiconductors, LCD displays, and, of course, watches. In this context, EMRoS (Epson Micro RObot System), the world's smallest self-propelled robot, is a positive example of technological interchange – from watch-making to robotics.

The idea of micro-robots long ago captured the imagination of scientists as well as writers of science fiction, who have envisaged roles for mini-automatons in medicine, spying and rescue missions, machine-servicing and space research. EMRoS, though, represents a new direction for micro-robots, which in the future could operate as communication partners, complete with personalities and behavioural ambiguities akin to those of humans. EMRoS is designed to stir emotional responses and sensibilities between technological creations and humans.

The EMRoS series, created by the 3M (Make Micro Mechanism) branch of Epson, made its prize-winning debut at the Japan Society for Precision Engineering (JSPE) symposium in 1991. In 1993, the first EMRoS, nicknamed 'Monsieur', was listed in the Guinness Book of Records as the world's smallest, self-propelled robot and was the object chosen to represent Japan in the time capsule at that year's Birmingham World Summit.

actual size

opposite and overleaf Epson EMRoS micro-robots

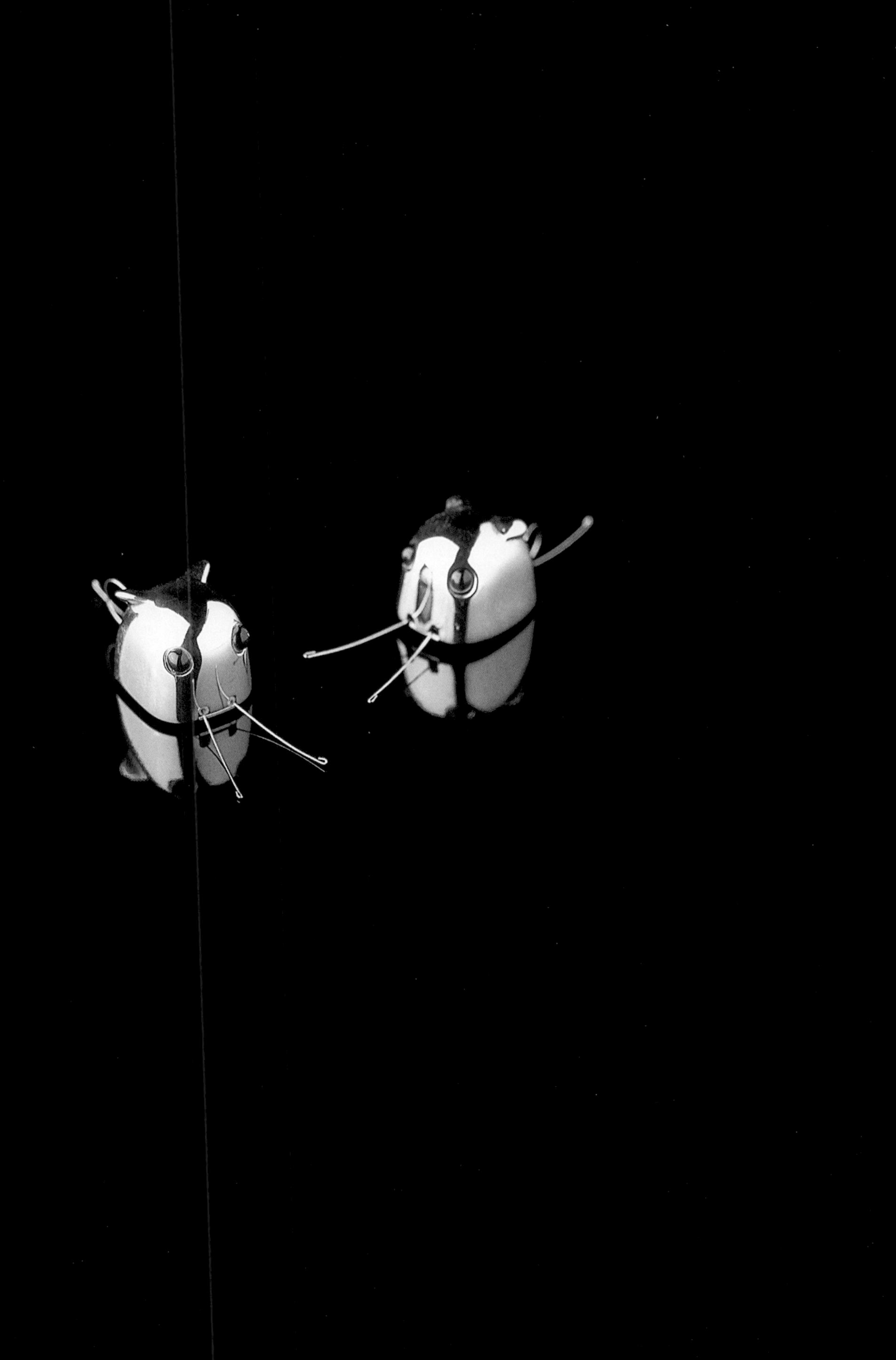

A tiny gem of engineering, the EMRoS has been designed and made to order in relatively small numbers since 1993. A diminutive rodent robot, with a propulsion volume of 1cc, 'Monsieur' uses watch technology - with 102 moving components, the equivalent of two wrist-watches - and is directed by sensors that are attracted by any light source, and propelled by its own integral power source.

The EMRoS has a heart of pure crystal - a precise, shock-resistant and ultra-small quartz crystal oscillator, which vibrates at a regular frequency when a small voltage is applied and emits small pulses to the drive unit. The power source is a rechargeable, high-capacity capacitor, again derived from watch technology, although in this instance, self-winding watches, where kinetic movement creates the energy for operation. The brain of EMRoS is a CPU-IC, a microcomputer, originally developed for multiple-dial analogue watches. The IC operates on very low voltages (0.003 Watts) and consumes little energy. Finally, all this micro-machine technology is controlled by phototransistor light sensors, which react in accordance with the strength of a light source.

Ironically, EMRoS is assembled by hand. Engineers at the EMRoS laboratories painstakingly construct models under a microscope, a task that requires all the skill and dexterity of a watchmaker. The possible uses for this kind of technology are limited only by the human imagination.

opposite EMRoS 'Monsieur'
below (left to right) EMRoS 'Lubie', Niño and 'Ricordo'
With a name that means 'recollection' or 'memory' in Italian, Ricordo remembers data when its sensor eye receives a stimulus. It can store up to 64 seconds of data, after which it reverts to instinct until prompted to repeat what it has remembered. To signify understanding of a command, its eyes light up. Ricordo does not forget a memory unless it is reset. Bearing a name that means 'caprice' or 'whim' in French, Lubie sulks, hesitates and makes capricious movements. It communicates via its illuminated eyes.

Fuji Photo Film Co., Ltd

Fujifilm F601 digital camera

With the image quality of consumer digital cameras rapidly approaching the holy grail of conventional film photography, Fujifilm has built an enviable reputation for medium- to high-performance in the relatively short period since this technology became available to the mass market. The flagship of the company's digital camera range is the Finepix F601, a model that embodies Fujifilm's unique 'vertical orientation' visual language, setting the controls and lens in a portrait rather than a landscape format. This design feature is central to the identity of high-end Fujifilm digital cameras, making the brand as instantly recognisable as the 'circle within a square' format from Canon.

The palm-sized Finepix F601 is the most recent camera in a line of digital convergence products where compatible, but usually separate, technologies are combined in a single product. Traditionally the problem with such products has been a perception that the quality of each function was inferior. The ISO1600-rated F601 avoids this potential pitfall by concentrating on features which are centred around the quality of the still digital image. As with most current digital cameras, apart from still image capture at a maximum of 2832 x 2128 (6.03 million) pixels, VGA digital video capture is also possible (at 640 x 480 pixels), as well as up to 272 minutes of digital audio recording, but it is the quality of the still image and the usability of the camera that comes through. These features, plus an autofocus 3X optical zoom lens, a multifunction user interface using a 38-mm (1.5-inch) LCD monitor, and a flash, are somehow squeezed into a tiny, chic aluminium package. In fact, the only restriction to reducing the camera to an even smaller package is the X3 optical lens, which demands a certain body thickness.

Photography has always been developed to follow characteristics of human eyesight, and digital photography is no exception. For instance, the human cornea processes light as an image on to the retina, which is made up of 126 million optical cells. It, in turn, sends nerve signals to the brain, which then identifies and translates them into what we describe as visual perception. For a digital camera the lens is the eye, the CCD (Charge Couple Device) is the retina and the signal processor is the brain. In the case of the Fujifilm CCD, there are other parallels. Nature, and the forces of gravity, mean that our world is generally based around verticals and horizontals rather than, say, diagonals. The unique Fujifilm CCD is

opposite Fujifilm F601 digital camera

actual size

designed to simulate the mosaic layout of the cells on the human retina by a 45-degree pattern angle of the honeycomb-shaped pixels, which arguably offers heightened sensitivity and clarity of image.

This kind of complex technology is difficult for most of us, as consumers, to appreciate fully: developments are about nuances and gradual improvements rather than radical revolution. However, this continual development of a particular technology, based around consumer requirements, has become a quintessential hallmark of contemporary Japanese industry, and represents the driving force behind technology-based products as we know them.

Casio Computer Co., Ltd

G-Shock watches

WQV-10 digital camera watch

Wearable technology, a catch-all term for any form of technology that is transportable and usable in a 'hands-free' way, including wrist-based products such as watches, is an expression of the desire to integrate technology with the user. Currently, wearable technology is superficial, in that it is attached to, or enclosed by the exterior of the body, but the future will see the everyday application of technology actually inside the human envelope.

For the moment, microprocessor technology offers us increasingly compact and convenient computing performance with three advantages – accessibility, affordability and wearability – resulting in the production of a range of watches which integrate a host of other functions, including mobile phones, GPS, compasses, televisions, cameras and MP3 players. The leading manufacturer of such portable products is Casio.

Outdoor pursuits such as yachting, climbing and skiing are ideally served by the Casio Pro Trek range, with models that incorporate a digital altimeter, barometer, real-time compass and thermometer, plus solar-power management and a tilt sensor (where the watch 'sleeps' until tilted towards the user to be read). The question for the designers is not so much about the technology itself, as how to create a usable interface which enables the wearer to access and exploit its potential to the full.

'Wrist networks', or watches with intercommunication capability, have been pioneered by Casio, including the world's first digital wrist camera, a wrist-worn PDA (personal digital assistant) and a wrist-worn MP3 audio player. The WQV-10 is a 2X digital zoom colour camera, enabling spontaneous image capture. It has a 1mb Flash memory storing up to 100 JPEG images at 16 million colours (on a PC) and can communicate its image to a PC or other WQVs via an infrared link. It can also tell the time.

opposite Casio Pro Trek multi-function watch

CASIO
TOUGH SOLAR
WR10BAR
MODE
COMP
BARO
ALTI
ALTI - BARO

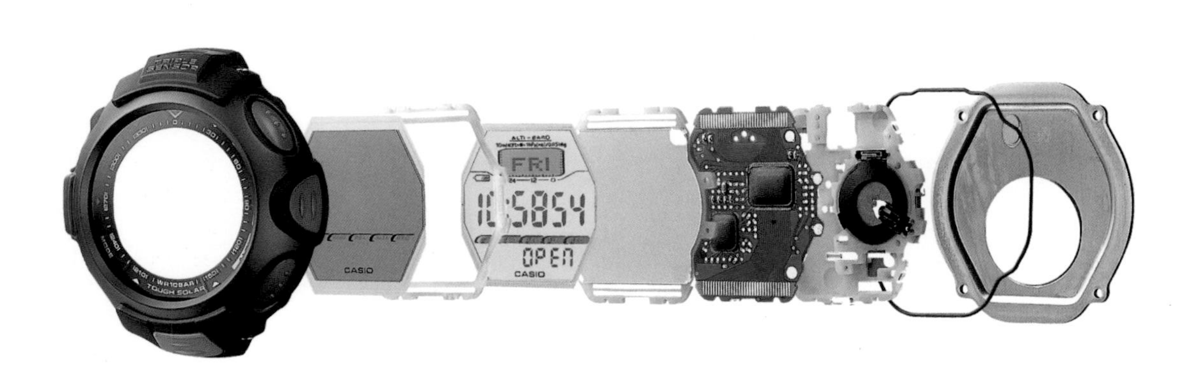
ALTI - BARO
FRI
OPEN
CASIO
TOUGH SOLAR

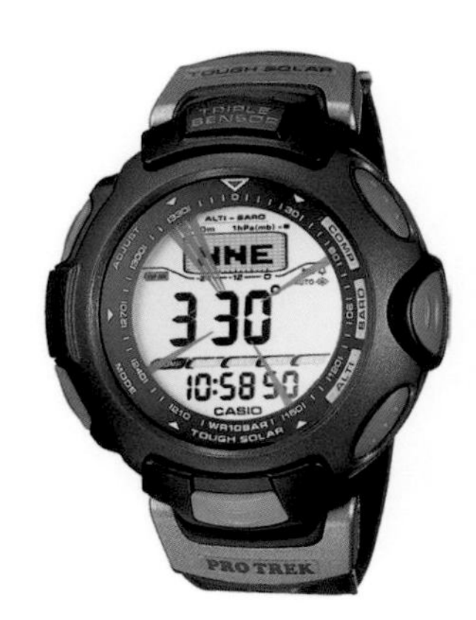
TOUGH SOLAR
CASIO
TOUGH SOLAR
PRO TREK

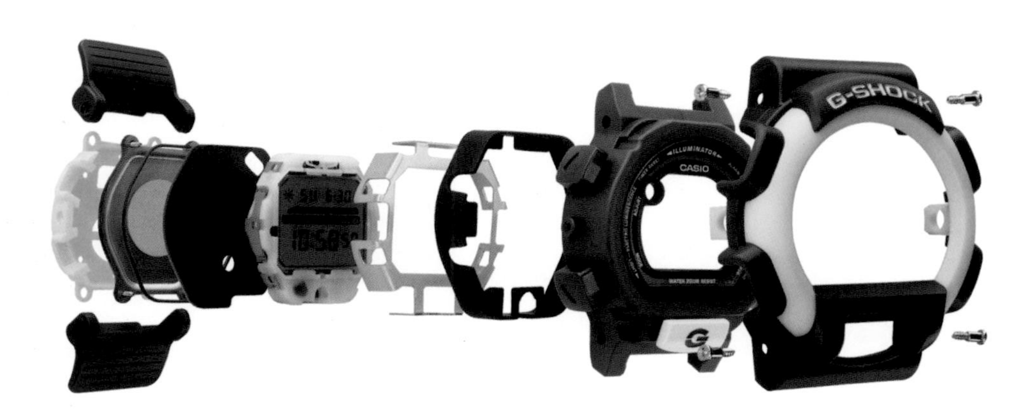
G-SHOCK
ILLUMINATOR
CASIO

CASIO
SPLIT-RESET
MODE / SELECT
WATER 100M RESIST
ADJUST
LIGHT
MODE
24HR
Baby-G

G-Shock watches

In the early 1980s, digital or LCD watches were at their lowest ebb in the consumer market, widely perceived as trashy, short-lived wrist trinkets. Analogue watches had reclaimed main-stream popularity, with companies like Swatch rising fast as a fashionable brand for younger consumers, producing stylish, desirable wrist-watches worldwide. Digital watches seemed to have been consigned to design history's 'round filing cabinet' along-side other, more recent, but transient techno-trends such as Tamagotchi.

But then, just as digital watches appeared doomed to obscurity, Casio's engineers made a technological leap forward with the creation of the Triple 10, a digital watch with a guaranteed ten-year life, that was water-resistant to a depth of ten atmospheres (100 metres or 109.4 yards) and that could withstand a drop of 10 metres (10.94 yards). It was the polar opposite of the restrained, compact analogue watches that Swatch was manufacturing in the early 1980s. Nonetheless, this 'toughest watch of all time', the so-called G-Shock (G = gravity), became one of the most successful watches, providing an excellent example of the value of continuous product development, as well as demonstrating a remarkable faith in a technology (LCDs) that was showing every sign of redundancy.

The Triple 10 technology involved the creation of a 'floating module', which is essentially the encasement of the core LCD component within a protective, shock-resistant and environmentally sealed urethane resin casing. This concept has been applied to all G-Shock models since 1983.

The G-Shock's toughness was acknowledged officially when the watch was adopted by US troops in the Gulf War in 1991. Images of the G-Shock in combat situations were widely broadcast, and the media coverage marked the beginning of a phenomenal rise in sales. The G-Shock was a natural complement to other fashionable items of clothing in the early 1990s, such as combat trousers and Catapillar boots, and high-performance materials such as Gore-tex and Kevlar. In 1995, Casio launched what was to become an even more successful iteration of the G-Shock, the Baby-G. Designed primarily for the 15- to 30-year-old female market, it turned into one of the biggest fashion trends of the late 1990s. Altogether, since 1983, 1.25 million G-Shocks and Baby-G's have been sold.

The constant reinvigoration of the G-Shock brand through the integration of adapted or emerging technologies – altimeters, thermometers, barometers, data banks, time-counters, GPS and MP3 have appeared in various models – is one of the most important elements in the continuing dynamism of the concept. Communication of the brand through sponsorship and advertising is also significant. For what is a very over-specified watch for most ordinary consumers, the positioning of the G-Shock as a lifestyle product has been valuable to Casio. Through advertising on youth-orientated channels such as MTV, special events sponsorship like snowboarding's 'G-Shock Air and Style' contest and celebrity endorsements, G-Shock has managed to create and sustain a unique credibility. Perhaps the most effective communication of the G-Shock brand has been achieved with the emergence of G-Factories, dedicated retail outlets in major cities which complete the G-Shock marketing culture.

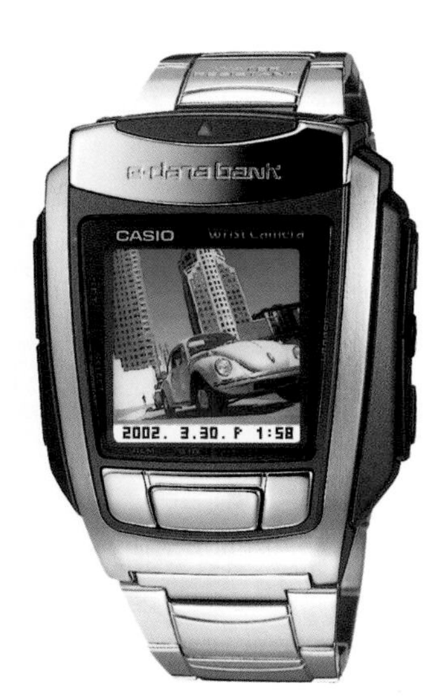

opposite exploded views of Pro Trek, G-Shock and Baby-G watches
left Casio WQV-10 digital camera watch

NEC Corporation

N502 i-mode Clamshell cellular phone

i-mode is the phenomenally successful Japanese cellular phone system from NTT DoCoMo (Japan's largest provider of mobile services). It has a share of over 65 per cent of the ever-expanding mobile communications market in Japan. Over half of the country's mobile handsets in use are internet-enabled, with users having access to over 10,000 internet sites developed specifically for mobile internet phones. Thanks to i-mode and its competitors, Japan accounts for 80 per cent of the world's mobile internet users, with i-mode alone having 23 million subscribers.

Cellular phones were first developed in the late 1980s with business users in mind. The introduction of digital mobile phones in the early 1990s, followed by a series of technological breakthroughs, allowed for increasingly light and compact handsets. By the end of the 1990s, cellular phone handsets had become multi-purpose gadgets for personal use. Their design also changed: from status symbol equipment to something more pocketable, fashionable and fun.

There are three basic designs for mobile phone handsets – the integrated type, the flip type and the fold-up type. The fold-up type, or 'clamshell' – where NEC leads – had a natural disadvantage in the early competition for greater thinness that characterised the market in the mid-1990s, in spite of the fact that it is compact when folded. The format also provides an appropriate distance between the ear and mouth when in use, making it easier to talk.

The development of the clamshell format in recent years represents vividly the behavioural differences between the Japanese mobile user and others. The typical mobile phone user in urban Japan will use an ear piece and a single thumb to scroll through sites and send messages, hardly ever raising the handset to the ear or taking their eyes from the colour LCD screen. Their preoccupation with internet data is possible due to the always-on nature of i-mode (packet-switched network), which means that users only pay for items downloaded rather than for the time spent online. Accessing and using this data is made easy by the design of the handsets.

opposite NEC N502i Clamshell i-mode mobile phone

Menu
HLD
2ABC
3DEF
4GHI
5JKL
6MNO
7PQRS
8TUV
9WXYZ

The conventional clamshell handset design has the keypad, the interface buttons and the single i-mode button (to display your customised menus) on the lower half, and the large colour LCD (liquid crystal display) on the upper, which flips open to take a call. Straps, bangles and other accessories all add to the sense of individuality and customisation.

The first clamshell handset, the NEC N502, which set a precedent for all mobiles in Japan, was designed by a man and woman design team, Toru Ichihashi and Chiaki Suzuki. It was originally designed for female users. Introduced in late 1999, it was a considered statement in simplicity and compactness. While other manufacturers produced what is termed 'stick phones', integrated, all-in-one handsets, NEC realised that the amount of complex analogue interface (button-pressing) and digital interface (data-control) activity suggested a logical division between the LCD and controls, and that the clamshell format was perfect for this control. But NEC's designers had to work hard to get the format approved by management. Early criticisms centred on the hinge as a potential weak spot, but customer-use testing made it clear that this had to be overcome as the reception was so positive. Problems with the hinge were solved by engineering a magnesium chassis which was strong and light, and which was made attractive by careful colour choice on the plastic casings.

The first LCD display on the N502 was monochrome. It was decided that it was preferable to have a larger black-and-white display than a smaller colour LCD, such was the amount of data that needed to be displayed. By choosing the largest display in the industry at that time (1999), NEC were able to successfully anticipate the popularity of internet usage on i-mode phones.

As with any designed object, a lot of user satisfaction is gained from an 'action' that communicates quality. The same kind of appreciation that is found in the pleasing 'thunk' of a car door closing, or the rotating action on the crown of a watch, is essential on a clamshell phone. Easy, single-handed opening, comfortable, error-free controls with good-sized buttons (especially important for those with long fingernails) and a firm button response are all essential details.

As technology advanced and component costs decreased, successors to the original model were developed. The N502i and N502iT were introduced, with enhanced features and functionality, including a porthole LCD on the cover (so a caller can be identified before answering), the industry's largest colour LCDs, compatibility with car navigation systems and physical reductions in size.

In March 2001, NEC released the N503i. It is an entertainment-orientated model which reflects the consumer-driven direction towards mobile entertainment as well as communication. Subsequently, the N503iS and N504i models have underlined this convergence. It is the premium feel, the simple, elegant finish that differentiates the design of NEC's cellular phone handsets from others in the market. It is easy to develop a different model by giving it a fantastic design. It is more difficult to develop a simple, high-quality product that everyone wants. The challenge for the future is how to introduce a new generation of phones capable of using the latest

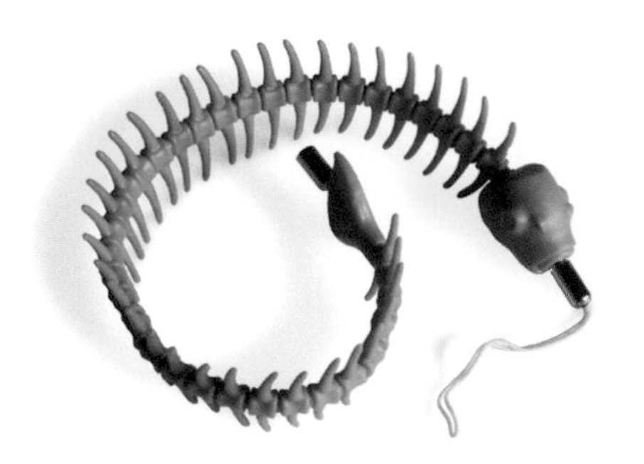

right generations of NEC i-mode phones
right bottom Na-cord phone strap by Maywa Denki
opposite N502i-N504i series of i-mode phones

3G (third-generation or FOMA in Japan) mobile phone technology, which will usher in video-enhanced messaging and more internet-based features, but with a marginal increase in the size of the handset. How the consumer will welcome an apparent step backward in size is difficult to predict.

SONY
RESET
SONY

5
consumer world champions

THE POWER OF SUCCESS

This chapter celebrates the demands of the singularly complex nature of Japanese consumers - one of the most switched-on, well-informed, fickle and interesting groups of buyers anywhere in the developed world. Japan has 126 million consumers in a land-mass roughly the size of the UK, a huge and vital market for Japanese companies, but one which is becoming increasingly fractured and unpredictable in its behaviour.

The products in this section are case studies in market success, either nationally or globally. All are based firmly on a long-term strategy, where the consumers' varying and evolving desires are addressed by the nature of the product itself. Customers' loyalty to, and familiarity with, these products is established by an investment in the communication of both brand and product excitement and desirability.

Examples include Sony PlayStation, clearly a product of its time, which has grown from Sony's corporate reputation for excellent entertainment products to epitomise the vigorous and dynamic nature of modern gaming. Nintendo's presence in the chapter confirms that there is room in this market for a different approach, based more on clear and imaginative characterisation in software generated in-house, and sustained by consistent marketing efforts, particularly in the transfer from one platform (Game Boy) to the next (GameCube).

Fujifilm's Cheki instant camera suggests that sometimes a new consumer market can be generated by the sensitive and creative development of an old technology into a fresh, exciting and fun product. This dual approach of inventive product development and fun styling is shown again by National's Electric Bucket washing machine and Soft Iron, products that soothe and entertain as well as function successfully. Finally we consider the unique conception of the WiLL brand, created by a disparate group of non-competing manufacturing companies, including Toyota, Matsushita, Asahi and Kokuyo, to sell lifestyle products targeted at a small but economically important consumer group.

opposite Sony PlayStation 2

Sony Corporation

PlayStation and PlayStation 2

Sony CCD-TR55 Handycam and DCR-IP7 Network Handycam

In 1994, a product was unleashed on the consumer world that was destined to become a cultural phenomenon. Even today it remains a benchmark in its field. Although the games market is now saturated with rival platforms, and some with better performance, it is the concept and image of PlayStation that continues to dominate the sector. Within three years this quintessential Japanese product had sold 10 million units worldwide; it has sold a further 70 million since. Like the names of many other Sony products – Walkman, Trinitron and MyFirstSony, PlayStation quickly entered the lexicon.

Sony had not had much success with video games prior to 1994, but PlayStation was an immediate winner. There were two major reasons for this: PlayStation took gaming out of the sweaty grip of the arcade devotee and placed the dynamic new product in the hands of home players, and Sony Computer Entertainment (SCE) encouraged a radical approach to, and involvement in, the PS platform by games creators, producing revolutionary titles such as *GranTurismo*, *Tomb Raider* and *Wipeout*.

The original purple-grey, 32-bit PlayStation console remains an icon as resonant as, say, the original Apple Macintosh computer. Both products captured the Zeitgeist, engaging users' imaginations in a way that their competitors could only dream about.

opposite Sony PlayStation 2

PlayStation 2
SONY
PlayStation
SELECT
START
ANALOG
MEMORY CARD

The PlayStation controller, the input device or physical interaction tool, was a *tour de force* of ergonomics and universally recognisable graphic iconography – the circle, the cross, the triangle and the square. Designed by Teiyu Goto, the classic original controller was replaced in 1997 by a dual-shock or vibrating version.

PlayStation 2 set new levels of realism in video gaming. Its processing speed gave games developers the opportunity to introduce previously unimagined levels of detail and dynamism, even by the standards of PS1. In addition, PS2 is a CD/DVD player with the Sony brand behind it and the capacity to act as an entertainment hub encompassing internet access for online gaming, television, shopping and other interactive applications. This kind of technical flexibility is crucial for products in this market, and although other platforms offer even better processing speeds and data-storage features, it is PlayStation that has seized the initiative – games developers use it as the basis for the development of new titles. And although PS2 was launched before its real potential was ready for exploitation, it has been able to capture a faithful following, as buyers know that compatibility with future technical developments will allow them to expand their PS2 as the mood – and technology – takes them.

right PlayStation
opposite PlayStation 2

PlayStation 2

CCD-TR55 Handycam

In 1989, Sony launched the TR55 Handycam. This passport-sized new product created a market share for Sony that has been sustained over the intervening years, with ever-smarter camcorders still bearing the Handycam name. The TR55 ushered in an era of camcorder products that were ergonomic, easy to understand and desirable to own, and in being so, set themselves apart from the complex, professionally derived analogue products available before. They also introduced Video 8, an 8mm digital format. This was the first time that users could capture the images they wanted without having to wield a heavy, bulky camera. Usability was further improved in later versions by the introduction of LCD screens.

DCR-IP7 Network Handycam

Over a decade later, Sony is producing the latest member of the Handycam family, the DCR-IP7; in its way as good a representative of the current state of audio-visual technology and usability as the TR55 was in 1989. The diminutive IP7 is really a network camera, in that it captures visual information digitally on tiny MicroMV cassettes and then transmits it in a compressed form (MPEG2) via a built-in modem. It can also transmit information wirelessly via bluetooth to any PC within a 10-metre (33-feet) radius.

These products, though they are generations of technology apart, share the success of Sony's brand qualities, so vital in this market sector, being compact, desirable, beautifully engineered and easy to use. These magical, creative tools are superb examples of how corporations like Sony have put powerful technology into the hands of ordinary consumers and taken them quietly into the future.

right Sony CCD-TR55 Handycam from 1989
opposite Sony DCR-IP7 Network Handycam

Carl Zeiss
Vario-Sonnar
SONY
PRECISION
LCD MONITOR
ネットワークメニュー
メール
アルバム
ブラウザ
設定
終了
SONY
SONY

Fuji Photo Film Co., Ltd

Fujifilm Instax Mini 20 (Cheki) instant camera

The onslaught of ever more affordable and effective digital cameras would suggest that chemical-based photography is destined for complete obscelescence. However, APS film was launched to postpone the inevitable demise of popular 35mm photography, and its equivalent in instant film photography, the 54 x 86mm (2.2 x 3 inch) sized 'Instax Mini' format film, for cameras like the Fujifilm Cheki, has saved the sector.

A staggering 1.5 million credit-card-sized, instant-print Fujifilm Cheki cameras have been sold worldwide since the product's launch in the mid-1990s. It is an excellent example of the development of what began as an obscure image format from the 1970s, and transformed into a successful, popular, low-cost, happy-snapper solution.

The Cheki is the latest version in this lineage of spontaneous, instant-image cameras, and represents one of three photographic sectors that Fujifilm has made its own – the others being disposable, single-use cameras and digital convergence products such as the combination of digital still camera, video camera and MP3.

While all cameras are, within reason, dependent for success on the quality of the images they produce, the affordability, convenience and image size have all contributed to the desirability of the Cheki format. This chunky, friendly, easy-to-use product incorporates a motorised f=60mm three-step macro lens, a flash, motorised image exit and an LCD display in a lightweight 310-gram (0.6-pound) body. It demonstrates how a previously niche product can be made universally attractive to a new group of consumers, for whom the immediacy and fun of instant image production outweigh the quality limitations inherent in instant cameras.

opposite Fujifilm Cheki instant camera

cheki
instax mini 20

Nintendo

Game Boy, Game Boy Advance

N64 and GameCube

Nintendo can trace its roots back to 1889 when Fusajiro Yamauchi, the great-grandfather of the current company president, manufactured Japanese playing cards called *Hanafuda*. This early enthusiasm for entertainment products has not waned over the decades, and Nintendo remains at the forefront of play-orientated technology. The phenomenally successful Pokémon collectables can be seen as a direct descendant of these early playing cards. However, Nintendo is one of the world's most recognisable electronic games brands and it is these landmark products that are of interest here.

In 1970 Nintendo took a brave step in introducing electronics into games for the first time in Japan. This led to a series of successful 16mm projected film-based games which were exported to amusement arcades the world over. However, in 1980, it was an in-house-developed character, originally named Jumpman, that really introduced the arcade world to the Nintendo name. This bestselling game, the rather obliquely named *Donkey Kong*, centred around a diminutive carpenter called Mario caught up in a race to save his girlfriend from a mad monkey. Studies have shown that Mario quickly became as familiar to children as Bugs Bunny or Mickey Mouse. In 1986 the Nintendo Entertainment System or NES (also known as Famicom) was launched as a domestic platform for these Nintendo characters and games. Other games soon followed, including *Tetris* in 1989.

Although Nintendo was not first in the queue at the start of the video-gaming revolution (arcade-based companies such as Atari, Bandai and Epoque got there in the 1970s), it was nonetheless instrumental in putting entertainment into the hands of millions of individual players when it launched the Game Boy. Even 20 years after its first appearance, this remains a landmark product, not just because of the number of units sold but also because of its creation of a cast of popular characters who have enchanted players worldwide.

opposite Nintendo GameCube

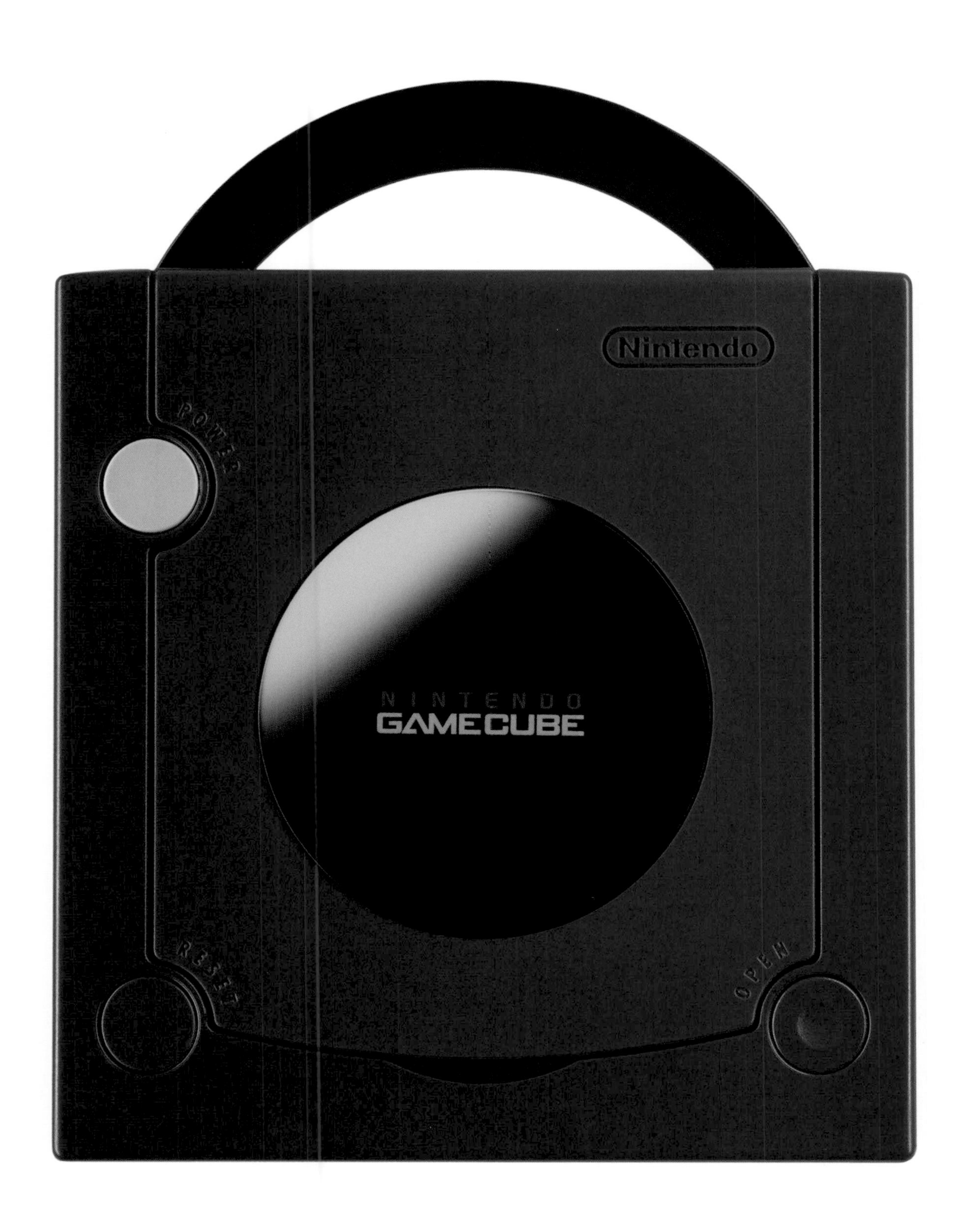
Nintendo
POWER
NINTENDO
GAMECUBE
RESET
OPEN

Nintendo's formula for success is founded on a pragmatic approach to capturing the imagination, particularly that of the younger player at whom classic games such as *Donkey Kong* and *Super Mario Brothers* were targeted. The company philosophy is to create and control its own software products, rather than carry externally-developed games like its swashbuckling competitors at Sony and Sega. Nintendo's approach is based on the assumption that gamers can be transported emotionally by a carefully crafted simulation of an imaginary world.

The Game Boy hand-held games console has sold an incredible 125.5 million units globally since its launch in 1989. Even now, its successor, the current Game Boy Advance (GBA), a more compact version of the original, still represents the bedrock of sales for the Nintendo company, which sold its one-billionth game cartridge in 1995.

Updating the Game Boy in 2001 from the boxy and rather unergonomic original involved more than an appearance change. To sustain sales in this fickle, young market, Nintendo needed to reinvigorate the Game Boy completely. So, the vertical arrangement was replaced by a landscape configuration and the basic, all-important user-interface controls that were so familiar to the generation that had grown up with the Game Boy were retained but with the addition of two shoulder buttons on the top of the GBA unit. The screen became a 2.9-inch TFT 240 x 160 LCD display driven by a 32-bit processor. Importantly, it was designed to be backward-compatible with previous Game Boys. Even more interestingly, the GBA was primed for future compatibility with converging technologies and ushered in the possibility of online gaming. The Game Boy Advance is currently the fastest selling console ever.

On 23 June 1996 Nintendo launched the N64 (appropriately named after its 64-bit processor, in theory twice as fast as Sony's 32-bit PlayStation 1). 500,000 units were sold on the first day, 2.3 million in the first year and 32.7 million since. Once again, the crucial interaction between the software and gamer was mediated by the controller. The family of Nintendo controllers, from that of the N64, the original (and some say archetypal) Super Famicon from 1988, to the current GameCube unit, are all as usable and effective as the controller of its direct competitor the PlayStation. Nintendo controllers all retain the basic icons and colours from Super Famicon, the familiar grey cruciform, four-way button and the

this page Mario and Pokémon game characters
opposite generations of Game Boy consoles

Nintendo®
GAME BOY ADVANCE

Nintendo®
GAME BOY ADVANCE

POWER
GAME BOY
Nintendo®
Nintendo
B
A
SELECT START

Nintendo®
GAME BOY ADVANCE

coloured A, B, X and Y control buttons. While the later N64 and GameCube controllers have evolved into more organic, ergonomic and three-dimensional shapes, to reflect the 3D nature of the games, the original controller shows how good hardware design can have longevity in an ever-changing software environment.

In autumn 2001, Nintendo launched the GameCube in Japan and the US. By the end of the year it had sold three million units. This is Nintendo's latest attempt to impose itself on a very crowded market against competing video-gaming platforms from Sony (PlayStation), Microsoft (XBox) and Sega (Dreamcast). For GameCube, Nintendo raided the IBM and NEC parts bin to create a 128-bit 405MHz chip to handle the processing, which results in approximately 14 million polygons per second performance (while PS2 works at around 3.5 million pps). The polygon is a flat plane, created in a 3D environment from three or more points which forms the basis or DNA of all video games, creating a tiled, faceted illusion of form in three dimensions. In theory, the more polygons processed per second, the higher the visual realism.

GameCube's games are delivered by a proprietary anti-pirate 80-mm (3-inch) DVD-derived disc, enough to store 1.5Gb of data. Shigeru Miyamoto, Nintendo's hardware designer, has created a further iteration of the familiar Nintendo visual language in the controller and drive, and a wireless controller is imminent. Up to now, the perception among players has been that Nintendo is aimed primarily at the younger end of the gaming market, an impression underlined by the game titles, and the slightly 'light' styling and colour choices of the unit. That will change as more games developers, and Nintendo themselves, develop more adult-orientated game titles (like the *Resident Evil* franchise) to take on competitors like PS2 and XBox.

But as the battle of the games platforms continues, and the emergence of online playing and do-it-yourself games programming challenges the very foundations of the gaming world, Nintendo is well placed to survive in this crowded arena because of its whole generation of populist and endearing characters. On this foundation, the company can build loyalty to the Nintendo name from an early stage in players' lives.

this page Nintendo GameCube
opposite Nintendo N64 consoles

NINTENDO64
NINTENDO64
NINTENDO64

Matsushita Electrical Industrial Co., Ltd

National NQ-SP10 'Soft Iron'

National NA-BK2 'Electric Bucket' washing machine

As part of this playful and colourful range of domestic appliances by Matsushita, the Soft Iron introduces a little joy and fun into the laundry process. Perhaps developed for leisurely ironing – when the play of form, material and colour can be appreciated – it represents a sympathetic and comfortable approach which has become a signature Japanese styling direction. The product is no gimmicky toy though; the Soft Iron uses a porous polyurethane and nylon sole that allows steam to pass through to the garment underneath, creating an efficient pressing effect. The styling, though, is the key differentiating factor, and it imbues the range with a humane, lighthearted but sophisticated touch.

The Electric Bucket is an aptly named small washing machine designed specifically for the Japanese domestic market, particularly compact urban apartments or even travel washing. By separating the motor and the clothes container, Matsushita has produced a simple, easy to use product that satisfies a real requirement for a miniature washing machine. The buckets can be bought singly and swapped over when the cycle is finished, ideal for small quantities of washing or specialist items requiring careful handling away from larger loads. The vertical, space-saving layout is ideal for tiny kitchens or hotel rooms, and the product concept is successfully focused on a real and defined need, and for that reason, is unmistakably Japanese.

opposite Soft Irons and Electric Buckets

National
NQ-SP10
National
NQ-SP10
National
NQ-SP10
タイマー
水流切換
National
NA-BK2
タイマー
水流切換
National
NA-BK2
水流切換
National
NA-BK2

WiLL

WiLL VS car: Toyota; KX-PW100CL phone/fax machine: Matsushita; WiLL tour culture: Kinki Nihon; stationery: Kokuyo; confectionery: Glica; cosmetics: Kao

WiLL is a unique joint branding venture between several well-known Japanese companies. It embraces a range of individual products manufactured independently by each participating company.

Asahi (brewers), Kao (household goods), Kinki Nihon (travel), Matsushita (consumer electronics), Toyota (cars), Glica (confectionery) and Kokuyo (furniture and stationery) all contribute a product of their choice to the venture, but share the development strategy and costs of marketing the brand. WiLL appears harmonious and complete in concept, while endorsing products that are disparate and individualistic in content.

Initially the brand was specifically targeted at younger consumers. The internet and 'guerrilla marketing' techniques (word-of-mouth, street flyers and events) were used extensively in the promotion of the concept, which is described as representing 'a joyful spirit and feeling of authenticity'. The promotion of any WiLL product by any of the companies involved, automatically promotes all the other products under the WiLL brand umbrella.

Branding is thought to have less value in specific markets in Japan than in other countries because of the tremendous range of products manufactured by the larger corporations. Matsushita, for instance, produces a huge array of products that cover almost every aspect of domestic life, making it difficult to encompass such diversity with a single branding message. Consequently, Matsushita operates a number of sub-brands – such as National and Panasonic – in specific market sectors. There are only a few companies, such as Sony, that invest in the promotion of their overarching corporate name and values, across all their product sectors.

opposite WiLL VS car from Toyota

WiLL VS
WiLL VS

Price, availability and reliability of products are key triggers for consumer choice the world over, but in Japan it is likely that the technical performance or function of competing brands will be very similar. This means that design has a huge role to play in selling a product and therefore it has been key in the development of the WiLL brand, with the goal of differentiating the products without reference to the parent company's profile.

The WiLL logo, an orange square with white typography, is the result of agreement between the different corporate members of the group and is applied to all the products under the brand umbrella, creating a true 'lifestyle' brand. However, industry observers outside Japan find it hard to believe that a consumer will faithfully keep to the WiLL brand: getting into a WiLL car to drive to a WiLL store for WiLL products; going on a WiLL holiday, where they will drink WiLL beer by the pool. The feeling of being targeted will surely be claustrophobic after a while. The reality is, of course, that although complete faithfulness to one brand is rare, many Japanese consumers are much less cynical about 'lifestyle' branding when it targets a specific group, which explains the phenomenon of the 'product craze' such as i-mode phones.

The progress of WiLL has been well documented both nationally and internationally since its launch in autumn 1999 and, despite a number of companies joining or leaving the group and the continuing domestic economic gloom, sales have been encouraging, particularly in the first two years of WiLL's existence. The initial three-year ¥600 million (US$5/£3.3 million) funding ran out in 2002, but the project continues apace nonetheless.

Several intriguing products have resulted from this exercise in corporate collaboration, most notably perhaps the WiLL VS car from Toyota, the instigator of the WiLL concept. Toyota was one of the first companies to realise that key consumer groups, notably the young, in the domestic market, had neutral or negative perceptions of large corporations and their products. WiLL was created to readdress those perceptions.

The website **willshop.com** is the virtual shopping experience that defines the WiLL philosophy and approach. Visitors can navigate around four 'shop levels' called Cool, Creative, Emotional and Relax, names which suggest the nature of the product promotions there. So, for instance, if you require a fridge, you can find it on two floors, Creative and Relax. A chocolate bar will be on Relax, and the WiLL VS car can be found on Cool and Emotional. There is a WiLL virtual café, various events based around the brand in major cities, and a genderless graphic style that has a sort of cool manga-ish look that works well in electronic format.

This exercise in collaboration and shared objectives has proved a commercial success – the WiLL beer from Asahi sold a million bottles a month after launch – but it is as much a conceptual triumph as a commercial one.

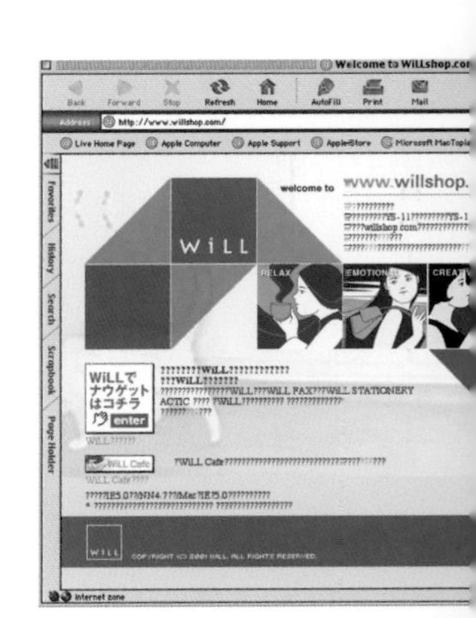

right WiLL confectionery and cosmetics
opposite (top to bottom) WiLL travel brochure, fax machine, website and staplers

WiLL
WiLL TOUR CULTURE
2001.11▶2002.3
和の体験 京都
in collaboration with 近畿日本ツーリスト

Panasonic

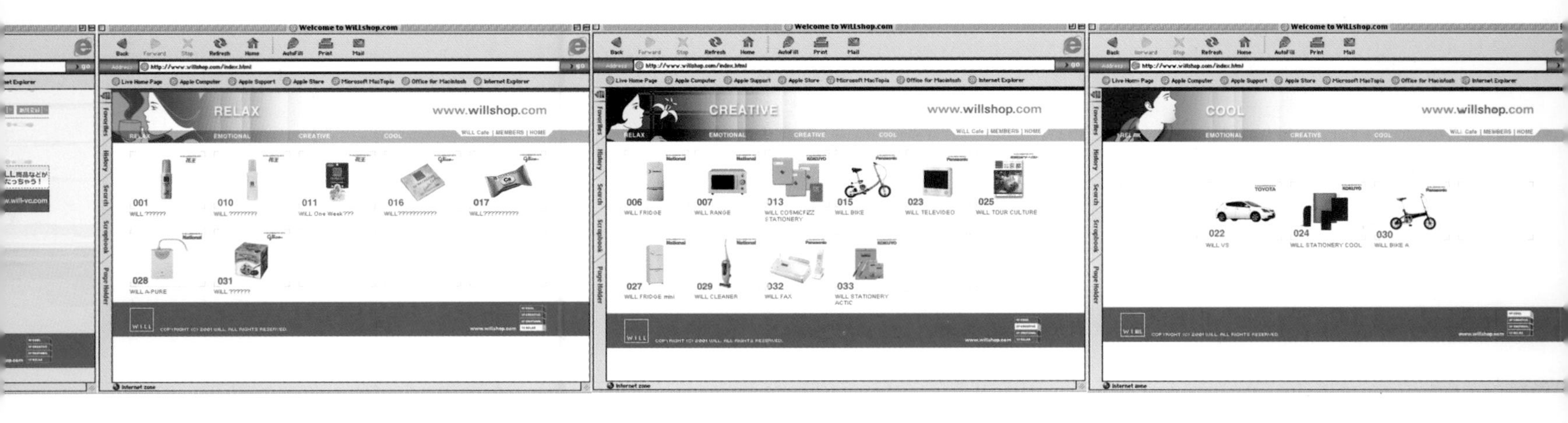
Welcome to WiLLshop.com
www.willshop.com
RELAX
CREATIVE
COOL

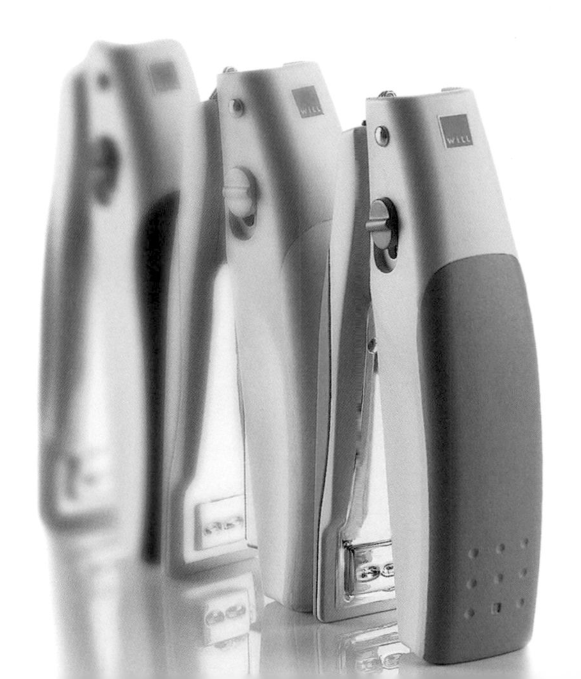

HONDA

6

transformations

EVOLVING NEW FORMS

Transformations is the story of traditional products, often manufactured using well-established, highly skilled artisan techniques, which have developed over time into examples of successful, world-class product design. These transformations of function and style can be seen in musical instruments, packaging, cameras, bicycles and even trains.

In this chapter, examples include Yoshikin's Global range of knives, developed from traditional craft techniques in the blade-making industry into highly valued, modern culinary tools and style icons. Yamaha's WX5 and WX7 MIDI products offer players the familiar interface of a traditional instrument from which to experience digital, multi-instrument playing in a cool and distinctive, contemporary form.

Fujifilm's single-use camera demonstrates how another disposable product – the result of a transformation from an expensive product to a throwaway one – can be safely and responsibly recycled and reused.

In transport design, many transformations grow from specialised applications in the professional sphere to clever solutions in the consumer world. Shimano's Nexave offers a taste of the efficiency, speed and ease of racing products to the leisure cycling market, and Honda's Caixa concept suggests a future where space and convenience replace aerodynamics as the chief consumer concerns in new automotive concepts.

opposite Honda Caixa Unibox concept car

Yamaha Corporation

WX7 and WX5 MIDI Controllers

Silent Taishogoto SH-30NR

In 1986 Yamaha's engineers, led by Yasuhiro Kira, produced the multi-award-winning WX7 MIDI Controller. A MIDI Controller (Musical Instrument Digital Interface is the standard used by virtually all electronic music devices) usually takes the form of a keyboard, but the WX7 demonstrated how innovative thinking can lead to new but still familiar forms of musical expression.

The WX7 and WX5 are designed to enable beginners and experts to play and control digital signals through a familiar, intuitive and ergonomic musical form, a woodwind instrument. It can be played to sound like a saxophone, trombone, flute or any other woodwind instrument one chooses. The WX5 also has different mouthpieces (single-reed or recorder) and four key layouts, which means that when a player selects a saxophone or a flute configuration, he or she will experience all the nuances particular to the layout of the real instruments.

The WX5 and WX7 are one of the most dynamic and expressive products of their kind, and generationally, the MIDI Controller has established a definitive aesthetic for the bridge between analogue and digital music creation. One of the most interesting results is that new players can be introduced to both woodwind instruments and digital technology at the same time, acknowledging the old, while embracing the new.

opposite Yamaha WX5 MIDI controller

WX5
YAMAHA

above WX7 prototypes
right WX7 and WX5 MIDI controllers

SH-30NR Silent Taishogoto

The rebirth of this traditional Japanese stringed instrument as a musical tool for the 21st century is one of the most fascinating examples of the transforming effect that technology can have on an established object.

The Taishogoto was first created during the Taisho era in the early part of the 20th century, and its beautiful and subtle harmonics are characteristic of classical Japanese music-making. In form it is a kind of mini-harp or zither which is placed on the lap or on a flat surface and played using either a plectrum, a bow or the fingers.

The Silent Taishogoto fits into the family of 'silent' instruments that have been so successful for Yamaha, enabling players to practise without disturbing neighbours, and to become confident in play without audibly making mistakes. Of course, the instrument can also be amplified and used for performance before an audience or in a recording studio. As with the other Silent acoustic instruments, the Taishogoto has no need of an acoustic chamber, which means that it is more compact and thus more transportable than traditional versions.

below Yamaha SH-30NR Silent Taishogoto

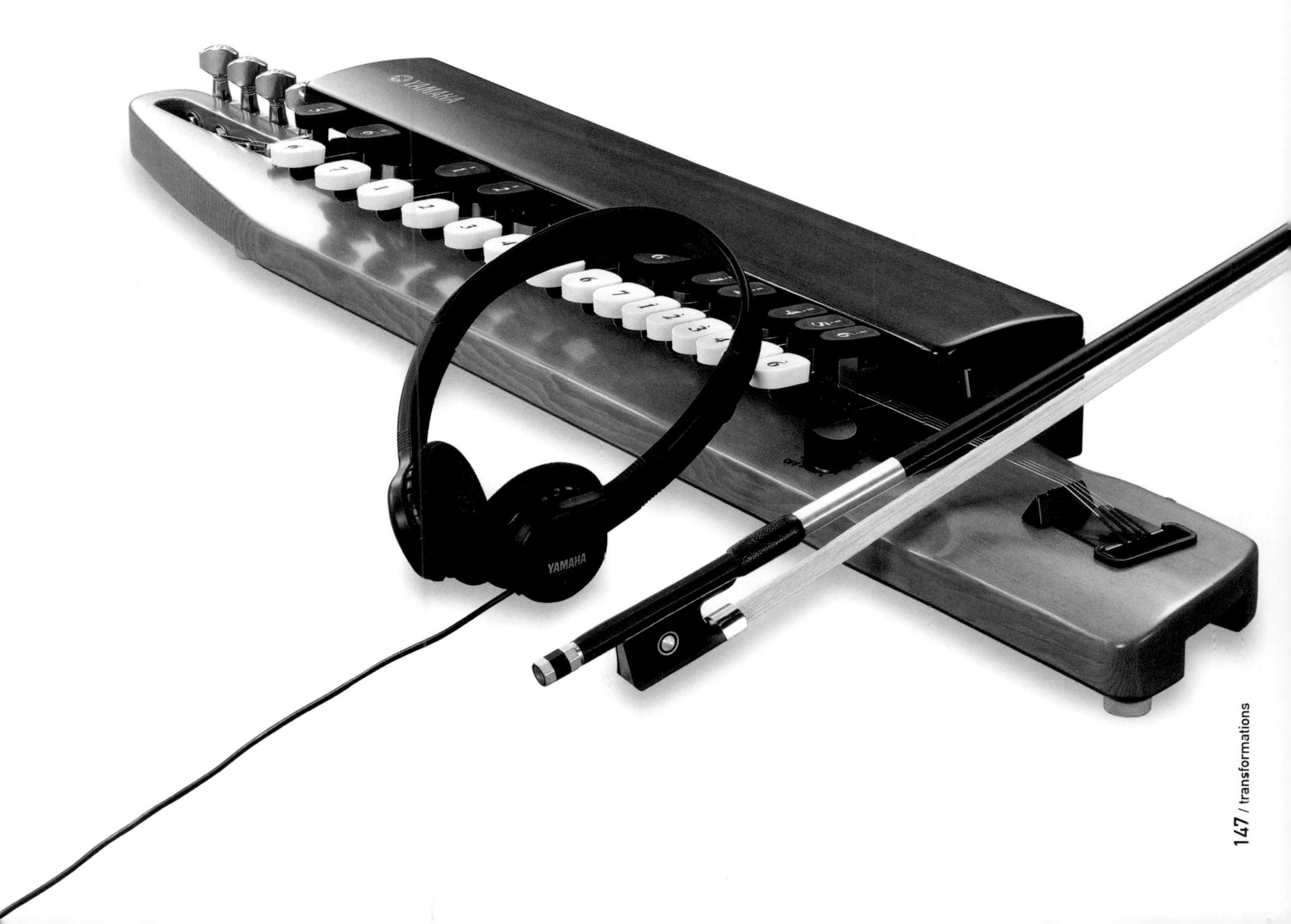

Yoshikin Metal Co., Ltd

Global kitchen knives

There are many things in life that are inextricably linked to a particular country or culture. When one thinks of swords, in the context of Japan, one's imagination leaps to evocative imagery of the Samurai, martial arts and Kurasawa films. Although not self-evident, the natural successor to the Japanese sword, or *katana*, is the modern kitchen knife. Japan is the home of food as an art form, and knives play a central role in its preparation and presentation.

In Japan, as in many other other nations immediately after the Second World War, food was scarce, forcing the defeated and exhausted nation to rely almost exclusively on its rice production for survival - hence the veneration of rice as a saviour of the nation even now in modern, prosperous Japan. In an island country where a third of the land is mountainous or unfarmable, fish also has a traditional place in the national cuisine. It is not surprising then that sushi and sashimi are such fundamental elements of Japanese food culture.

In many ways, the Global range of knives by Yoshikin, a company based in Nigata prefecture in northern Honshu, has come to epitomise the Japanese ability to transform a traditional, highly specialised tool into a successful generic product that can be exported worldwide. Whether we buy such cutting tools for purely functional requirements or because we want to acquire some highly attractive kitchen accessories (or indeed both), this range represents something very progressive in its field, a fact recognised by professional and amateur chefs worldwide, and a collection of international design awards.

opposite Yoshikin Global kitchen knives

GLOBAL
CROMOVA 18 Stainless Steel
GS-3 YOSHIKIN Japan
GLOBAL
CROMOVA 18 Stainless Steel
G-2 YOSHIKIN Japan
GLOBAL
CROMOVA 18 Stainless Steel
G-4 YOSHIKIN Japan
GLOBAL
CROMOVA 18 Stainless Steel
GS-3 YOSHIKIN Japan

The knives were designed in the late 1980s by Komin Yamada. Unusually for Japanese products, their appearance has remained virtually unchanged for over ten years, a demonstration perhaps of the initial 'rightness' of the design. When creating the Global range, Yamada says he tried to avoid the conformity that he perceived in traditional Japanese knife design, although ironically, many people outside Japan consider the knives to be unmistakably Japanese in form. From the fearful 320-mm (12.5-inch) long GP10 round-ended sword-like *bashi* to the diminutive stiletto-like GS8 gutting knife, these knives represent the high-point of performance and presentation. In the years since their launch, the range has been broadened to include a giddying array of cutting kit for the kitchen, with 40 different knives for fish and fowl currently in the Global and Global Pro ranges.

Over 800,000 Global knives have been manufactured since 1983, and up to 85 per cent of production is for export to 15 different countries. Denmark, Sweden and the UK are the biggest markets, with Japan, surprisingly, only in sixth place in terms of sales. These lethal knives are crafted in molybdenum vanadium steel, a material usually found in surgical knives, and are as balanced in the hand as they are effective in cutting. The blade is made more durable by a thermo-treatment called annealing. This process involves heating the steel to a very high temperature to achieve an 'austenitic' state, and then cooling it rapidly so that it reaches a very hard 'martensitic' state.

There are eight stages of production for a Global knife:

1. The raw stainless-steel sheet is stamped to achieve a basic blade profile.

2. The basic blade profile is heat-treated.

3. The two halves of the handle are stamped and welded together.

4. The knife blade and handle are then welded together.

5. The handle is black chromium-plated, which is then ground to reveal the black filled dimples on the handle.

6. The blade is roughly ground and finished automatically.

7. The Global motif is etched on the blade.

8. The handle is polished and the blade finally sharpened.

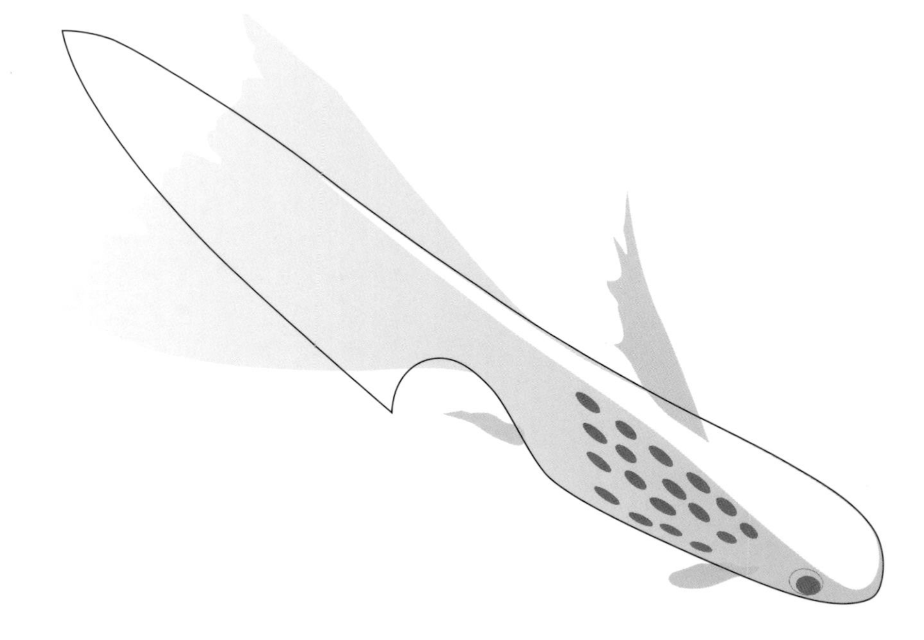

right Komin Yamada's concept sketch for Global knives
opposite Yoshikin's traditional Yanagi and Fish Deba knives

GLOBAL
CROMOVA 18 Stainless Steel
GS-3 YOSHIKIN Japan
GLOBAL
CROMOVA 18 Stainless Steel
G-2 YOSHIKIN Japan
GLOBAL
CROMOVA 18 Stainless Steel
G-4 YOSHIKIN Japan
GLOBAL
CROMOVA 18 Stainless Steel
GS-3 YOSHIKIN Japan

The knives were designed in the late 1980s by Komin Yamada. Unusually for Japanese products, their appearance has remained virtually unchanged for over ten years, a demonstration perhaps of the initial 'rightness' of the design. When creating the Global range, Yamada says he tried to avoid the conformity that he perceived in traditional Japanese knife design, although ironically, many people outside Japan consider the knives to be unmistakably Japanese in form. From the fearful 320-mm (12.5-inch) long GP10 round-ended sword-like *bashi* to the diminutive stiletto-like GS8 gutting knife, these knives represent the high-point of performance and presentation. In the years since their launch, the range has been broadened to include a giddying array of cutting kit for the kitchen, with 40 different knives for fish and fowl currently in the Global and Global Pro ranges.

Over 800,000 Global knives have been manufactured since 1983, and up to 85 per cent of production is for export to 15 different countries. Denmark, Sweden and the UK are the biggest markets, with Japan, surprisingly, only in sixth place in terms of sales. These lethal knives are crafted in molybdenum vanadium steel, a material usually found in surgical knives, and are as balanced in the hand as they are effective in cutting. The blade is made more durable by a thermo-treatment called annealing. This process involves heating the steel to a very high temperature to achieve an 'austenitic' state, and then cooling it rapidly so that it reaches a very hard 'martensitic' state.

There are eight stages of production for a Global knife:

1. The raw stainless-steel sheet is stamped to achieve a basic blade profile.

2. The basic blade profile is heat-treated.

3. The two halves of the handle are stamped and welded together.

4. The knife blade and handle are then welded together.

5. The handle is black chromium-plated, which is then ground to reveal the black filled dimples on the handle.

6. The blade is roughly ground and finished automatically.

7. The Global motif is etched on the blade.

8. The handle is polished and the blade finally sharpened.

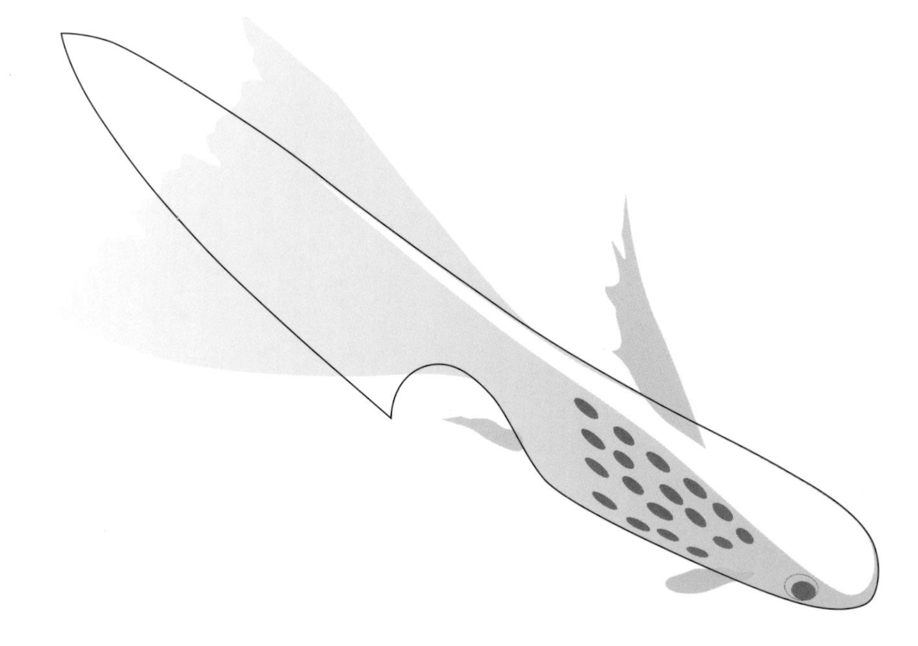

right Komin Yamada's concept sketch for Global knives
opposite Yoshikin's traditional Yanagi and Fish Deba knives

Yoshikin also produces an equally commanding range of traditional wooden-handled knives. In manufacture and appearance these beautiful blades are more closely related to the traditional techniques of knife- and sword-making. The *Yanagi* sashimi knife blade is one-sided, so that as it glides through the fish, the thin slices of sashimi fall away from the blade in a glistening arc. To dissect, bone and fillet fish, one uses the *Fish Deba*, a single-sided ground blade that has been perfected over centuries to achieve the swift and accurate removal of bone from flesh. Maintenance of blades like these is crucial. The essential requirement of any chef's knife is the acuteness of its blade, and its upkeep is the responsibility of the owner. With both these ranges, the knives are at the mercy of their user, and beg to be kept honed to perfection. In return, they will remain faithful and reliable assistants, like the sword to its master in history.

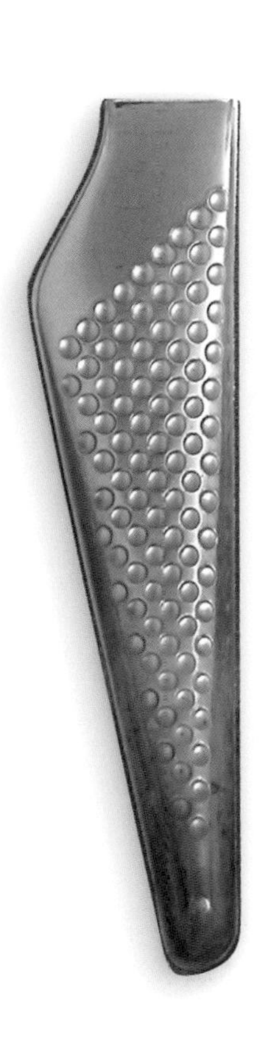

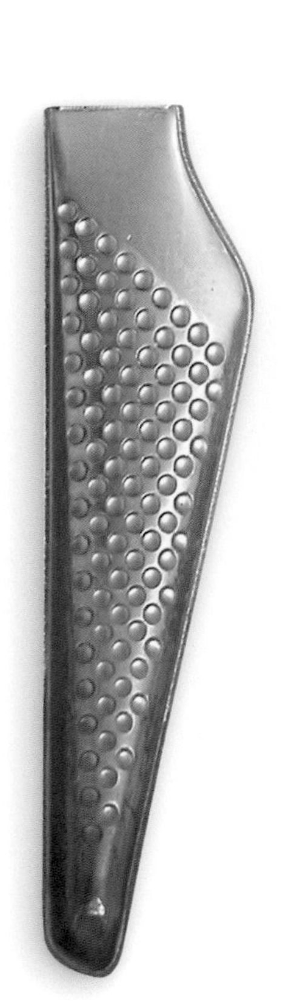

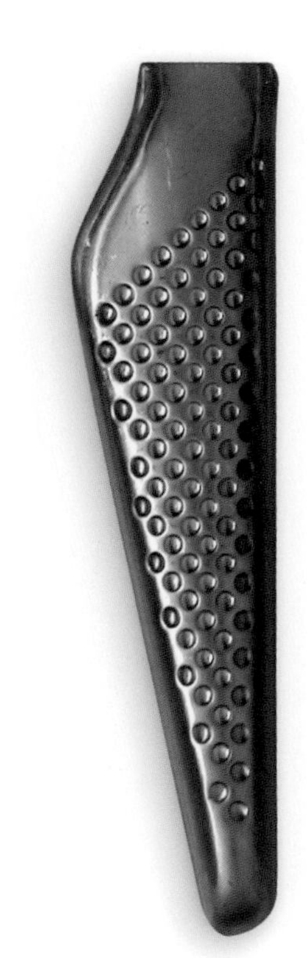

stage 1

stamping to achieve profile

stage 2

heat-treating of profile

stage 3

stamping and welding of handle

stage 4

joining of blade and handle

stage 5

black-chromium plating

stage 6

grinding and finishing

stage 7

etching and sharpening

stage 8

final polishing and sharpening

Fuji Photo Film Co., Ltd

Fujifilm 'film with lens' single-use camera

The disposable camera is one of the finest examples of cost-driven, user-orientated functional engineering. The purpose of such cameras is to take images of an acceptable quality, to be inexpensive to buy and intuitively easy to use. In addition, they are manufactured to be readily disassembled and recycled. In fact, in Japan they are called 'films with lenses' (FWLs), a good description of the ultimate goal of the FWL engineers: to make a roll of film virtually take its own pictures.

The disposable camera is all about macro-economics and micro-engineering. Every one of the FWL's 120 components has a life-span just long enough to carry out its purpose, and as a disposable product, the FWL can have a hefty ecological impact on the planet after use. The film emulsion process creates a high level of pollution; the waste and pollution involved in the disposal of plastics is even greater. This ecological burden would have been extensive without design for recycling or reuse, as a staggering 250 million FWL's have been sold since 1986, 70 million in Japan alone.

There is a real distinction between the terms recyclable and recycled, between the potential and the actual. In this case, the operative word is recycled, a positive reality. Fujifilm's inverse manufacturing facility uses the same trucks that deliver new and unused FWLs to the shops to return used products to the factory for recycling.

opposite Fujifilm film with lens/disposable camera

SuperEye
800
FUJIFILM
ON
OFF
ON
OFF
フラッシュ
スイッチ▶
FILM WITH LENS
ちきゅうにやさしい
プラスチックの再利用

The FWL has eight basic components. The front and rear body, the switch unit and the advance wheel are all polystyrene and can be reground and remoulded. The battery, main unit, flash unit and lens, on the other hand, can simply be checked and cleaned before reuse. Even the card outer sleeves are recycled. This type of reuse has to be planned meticulously at every stage, from the design and manufacture of the product to the collection of used products for reprocessing. Fujifilm's factory in Ashigara is a leading-edge facility for this purpose.

The plastics engineering needed to produce such a complex yet inexpensive product is extraordinary. All parts on the FWL are snap-fits, which means that disassembly can be swiftly carried out by automated laser-guided machines that identify and dismantle the different camera components. All lenses, packaging, plastics and electronic components are reused and the whole process takes less than a minute.

above Fujifilm film with lens assembled

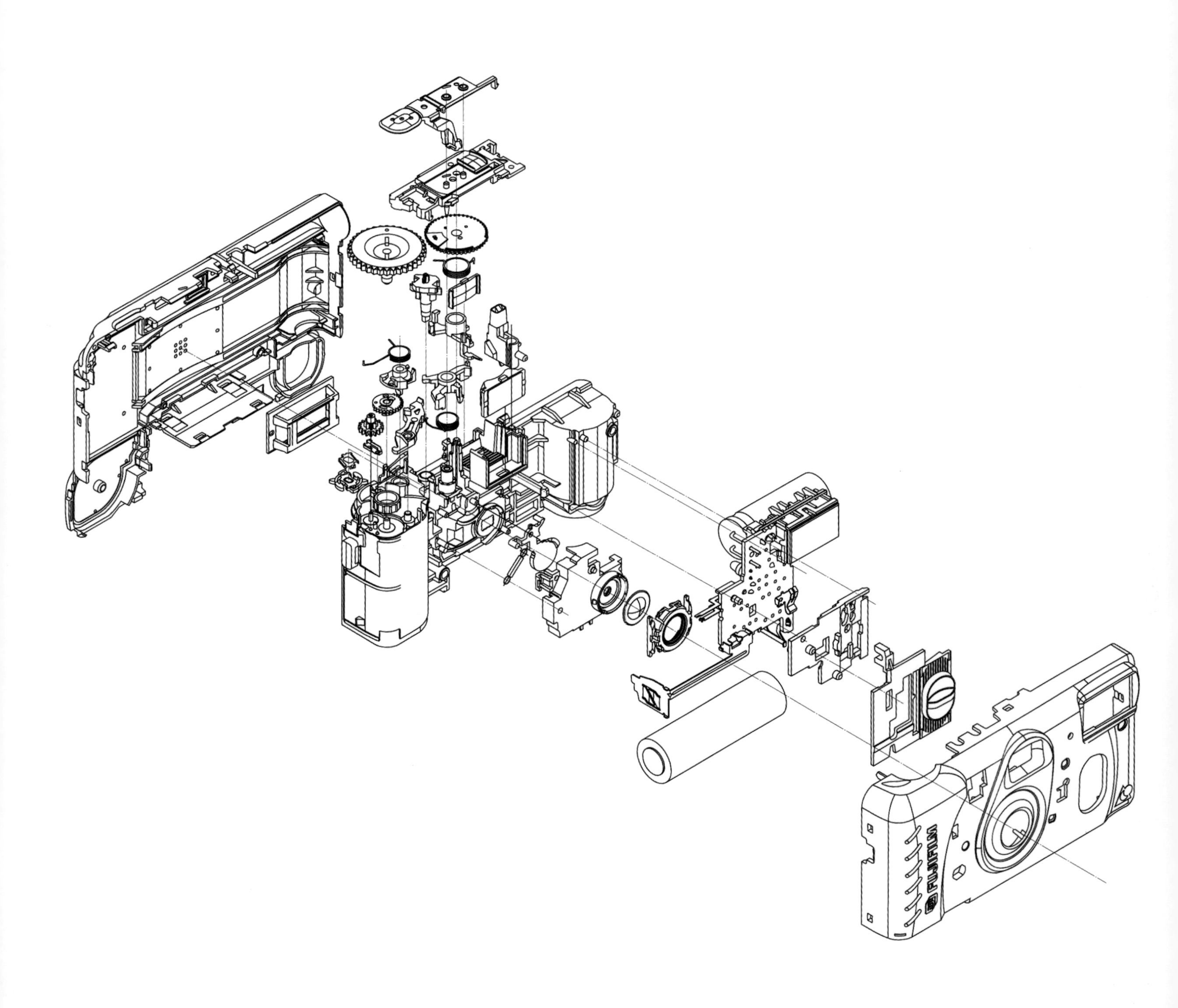

above exploded CAD view of components
opposite breakdown of components for recycling and reuse

FUJIFILM
39
F
CN-16

Shimano Inc.

Nexave C910 computer-controlled component system

The Shimano Nexave system, or Digital Integrated Intelligence (Di2), is a fascinating development in leisure cycling. This technology gives the rider the simplicity and comfort of integrated automatic gear-shifting and suspension control, achieved by digital and electro-mechanical means.

The rider selects one of four modes depending on environment or mood: automatic gear-shifting for the best balance of speed versus pedalling ease; a fast mode for speed cycling; a leisure mode for easier pedalling; or a manual mode for selecting gears at the press of a button. A system of gliding sprockets means all gear changes go almost unnoticed by the rider. While this is going on, the Nexave computer adjusts the hydraulic front and rear air-suspension dampening, to achieve the most efficient balance of power and comfort. When a rider is setting off or climbing a hill in a lower gear, the control unit will automatically select a hard suspension setting for maximum efficiency. A 'Flight Deck' control panel keeps the rider informed about the gear setting, speed, and so on.

While manual gear shifting is not the most taxing activity in the world, this development maximises the use of the kinetic energy produced by the rider, and allows them to concentrate on the scenery or where that bus is going, in relative comfort. While the added weight of such a system might be counter-productive on a racing bike, on a commuter model it is the sort of thinking that will be appreciated by the majority of riders.

Shimano's objective with this technology was to enhance the riding experience rather than just maximise speed and efficiency – as they have done with other 'sports' components, such as the remarkable Dura-ace. The future will bring iterations of the Nexave which are lighter and even more efficient, probably making it as ubiquitous as other Shimano bicycle components, particularly in Japan where commuter cycling is at its most prevalent and organised.

opposite Shimano 3.3.3 Freewheel from 1921

3.3.3. FREE WHEEL
MADE IN JAPAN TRADE MARK

above Shimano Dura-Ace deraillier
opposite top Nexave computer controlled deraillier
opposite bottom Shimano's fully Nexave equipped concept bike

SHIMANO
NEXAVE

CONCEPT

Honda Motor Co., Ltd

Caixa Unibox concept car

The Japanese car industry has always invested heavily in R&D. It is true that there have been examples of soulless, mass-produced cars that failed to capture drivers' imaginations, but there have also been instances of inspired and quirky styling, like the Nissan S-Cargo van and the Figaro cabriolet of 1989. The S-Cargo was a cleverly styled homage to the venerable Citroen 2CV; the Figaro, a nostalgic cartoon sports car of the 1950s. Today, when the Japanese motor industry gets it right, the whole world sits up and takes notice. Honda has a rich history of innovation and success on both road and track, and is rightly known for its engineering virtuosity. The Caixa Unibox demonstrates the ways in which the company's designers are now responding to the modern challenges of an increasingly congested road system.

Tokyo and Osaka are the nearest we have to the futuristic metropolises that literature and film have long promised us. The Caixa Unibox is a vehicle that has been designed for just such a crowded environment, where maximising interior space is as important as engaging the imagination of the driver. The lateral creative thinking displayed by the Unibox demonstrates a new attitude and approach by contemporary Japanese automotive designers, working in a more unrestricted creative environment.

The Unibox is designed to give drivers and passengers the comfort and flexibility that are extremely valuable when low speeds make rakish aerodynamics far less important than space maximisation. The car's exterior panels are modular, suspended from an integral truss frame, and allow owners to customise the appearance and function of their vehicles. An expansive built-in storage system is integrated into the door cavities. These are spacious enough to hold a mini bike for when traffic jams get really bad, and a generator to power a motorised shopping trolley. The tall, vertical sides of the Unibox give passengers the feeling of being in a light, spacious room on wheels. The interior has been enhanced with wood strip flooring, the use of leather and other natural materials, and a flexible seat layout.

opposite Honda Caixa Unibox concept car

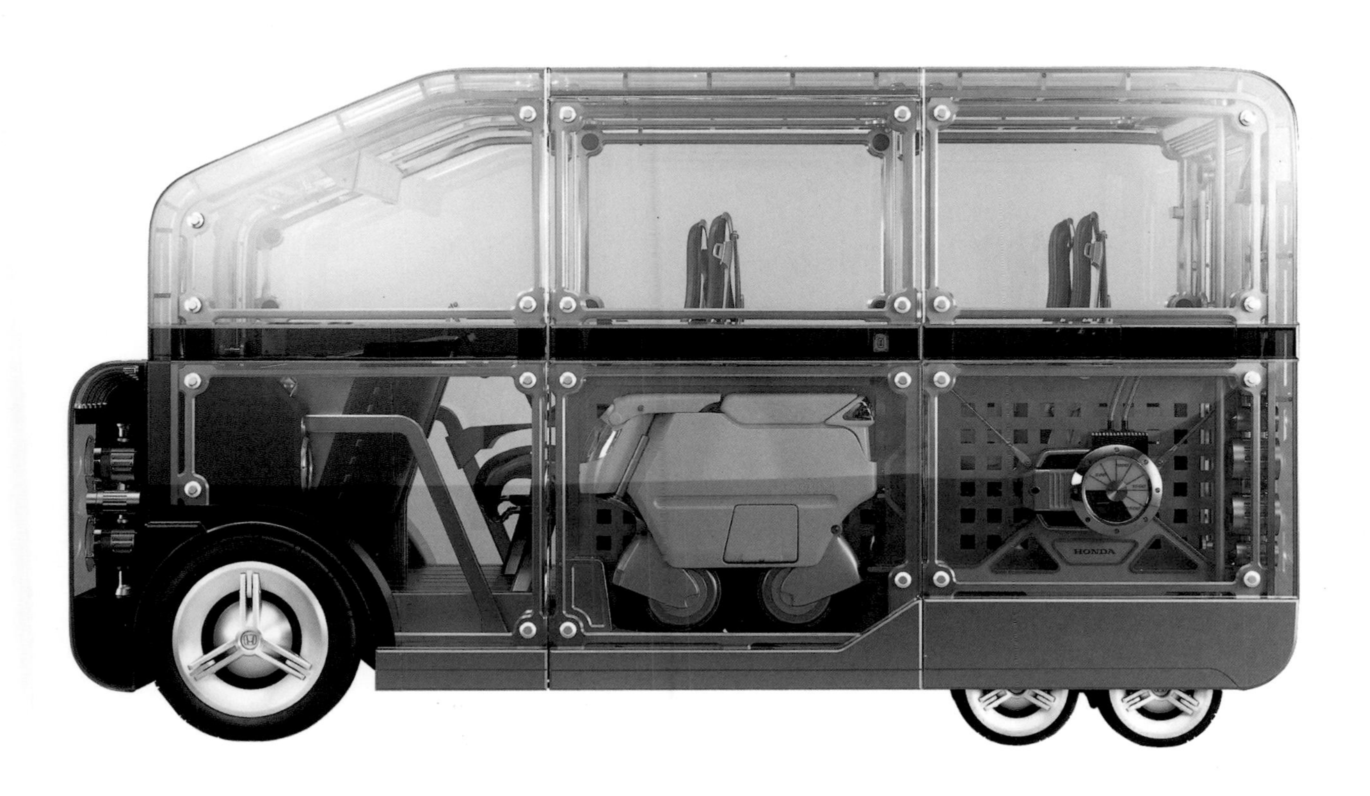

Instead of a steering wheel, the Unibox has a joystick – a concession to younger consumers more familiar with video games than cars. Technology makes its presence felt by the addition of milliwave radar and CCD cameras, designed to communicate with other vehicles to avoid collisions and obstacles. The rearview mirror has been replaced by cameras, which give a 180-degree view in a long LCD panel. This car may well be the shape of things to come.

right the Caixa mini-bike is stored in the door cavity (top) and design details of the Caixa Unibox (bottom)

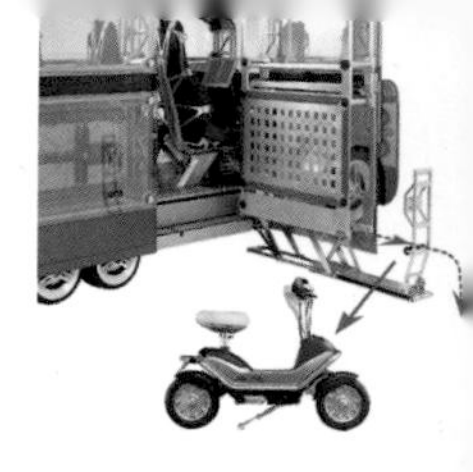

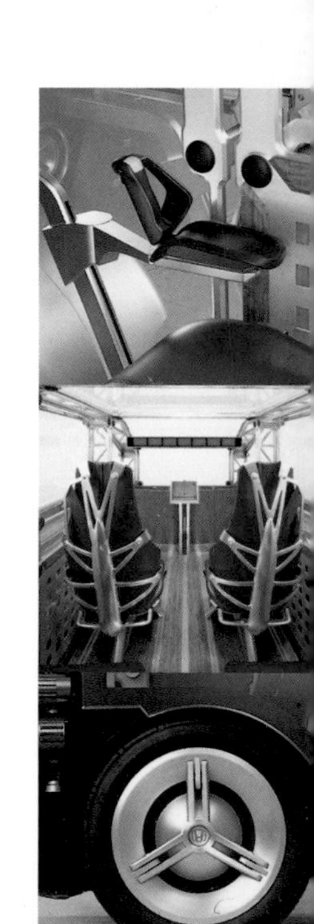

above left Sendai Mediatheque designed by Toyo Ito
above right Issey Miyake's A-Poc store in Osaka

new japanese design

YUICHI YAMADA, CHIEF EDITOR OF *DESIGN NEWS*

Until recently, Japanese designers, especially industrial designers, were born out of corporate design departments. The strong organisational systems of these design departments gave power to their companies against global competitors. Companies, such as Sony, Honda and Toyota, came to represent what Made In Japan really means: top quality and design excellence. Japanese industrial products, created by each company's design team, took the lead and dominated world markets. The activities of the corporations' design departments focused on the merits of mass-production, rather than ways of differentiating their products from those of their competitors. The result was uniformity and a lack of ground-breaking design.

To better the situation, companies restructured their design departments, introducing a process of open-ended design, which has a broad connection with the consumer. In the last few years, Japanese industrial design has undergone a transition, changing significantly both company activity and society. This reorganisation of industry's design sectors, and the creation of subsidiaries within existing companies, is considered remarkable. For instance, in the last two years, Hitachi, Sanyo and Matsushita, the market leaders in white goods and domestic product manufacturing, restructured their design groups, one after another. Hitachi Ltd, Design Division, Sanyo Design Centre Co. Ltd and Panasonic Design Company have all created subsidiaries, in an attempt to reorganise their design teams.

Despite the manufacturers' design departments having been established over a long period of time, from 1950 until the 1980s, the majority were only restructured in the late 1990s. The background to this was the bursting of the economic bubble in the early 1980s and the subsequent economic decline, which continues to this day. In order to overcome such chilly economic conditions, a survival strategy has emerged, which changes the designers' role, allowing them to create the next generation of products. Companies are using brand identification as a weapon. The need to strengthen the human interface of products and to address the increase in skills, such as IT, is creating a new movement in industrial design.

So, how can we create world-class products in Japan that aren't informed solely by marketing thought or current product development trends? The answer could lie in the rise of the *Dokuritsu-kei* designer.

The recent work of young, thirty-something freelance designers (*Dokuritsu-kei*) has been remarkably different to that of the company designer, whose products used to be the mainstay of Japanese design. The *Dokuritsu-kei* have strong opinions. They are fighting for something every day. From a different perspective, they are outsiders who are committing heresy by deliberately occupying the margins of ordinary design. They have experience of working for the big corporations, have travelled overseas and worked as freelance designers, and, as a result, have developed their own temperament. Not influenced by anything, they create their own individuality and power of expression.

Ichiro Iwasaki has established a new audio brand Mutech with a Korean manufacturer. In Korea, product design and the management of freelance designers is more advanced than in many Japanese companies. Masamichi Katayama, who designed the Marc Jacobs and A Bathing Ape stores, has achieved great popularity as a designer, earning the title *Buppan no Katayama* (Mr Shop Design). Katayama has developed two remarkable companies, Anotherwall, which focuses on product development, and Wonderwall, which mainly designs commercial interior spaces. Shin Nishibori stepped into the limelight when he designed the CD player/radio P-case. He went freelance after leaving the Matsushita in-house team and has recently joined the design team at Apple Computer. Tokujin Yoshioka experimented with new industrial materials for client A-Poc at its Osaka shop, and has created exhibitions for Audi and the chair concept Honey-pop. His bold design expression and experimental style have earned him praise, not only in Japan, but also around the world, and his work is now very much in demand.

The approaches taken by these designers were made possible through their independence, individuality and original design strategies. How can large corporate clients incorporate these strong, individual styles? The company design department, traditionally responsible for project management, needs to integrate freelance designers and their way of working carefully into the organisation, in order for Japanese design to become broader and more innovative.

The key to Japan's design future

The last three winners of the Japanese Good Design Award (Grand Prix) were the Aibo entertainment robot in 1999/2000, the A-Poc fashion-product concept in 2000/2001 and the Sendai Mediatheque public architecture design in 2001/2002. These designs all have their own character and are notably different from most products that had been made or developed in Japan previously.

Sony's Aibo is not a helping-human product. It does not replace the function of part of the human body, like a washing machine (arms) or car (feet), making a traditional task easier. This robot dog is a product which doesn't give humans any advantage in terms of ability. Rather, it is a pet, which can heal the human spirit and take the role of a communications partner.

Issey Miyake's A-Poc creates a new relationship between man and fabric. Derived from the phrase A Piece of Cloth, the product can be adapted to be a fashion garment or a piece of furniture. The consumer can remove the structure and wear the cover. Because the designer controls everything, from the weaving of the fabric to the production process of the finished item, it is

right original Sony Aibo ERS-110

possible to reduce manufacturing waste and to allow production to respond to individual orders. It is a prediction of what production could be like in the future, meeting the contradictory aims of individually tailored goods and quantity manufacturing.

Toyo Ito's plans for Sendai Mediatheque changed the concept of what public architectural space should be. The structural principles of computer programming were used to create a building constructed along similar lines. It is not only the physical replacement of beam and pillar with high-tech plate and tube that is important, but also the way the building functions. The model of the traditional museum and library was reworked to create flexible information spaces for a variety of activities – reading a book, using a computer, watching a film, viewing an exhibition, producing digital works, participating in workshops and communicating with other people. This is architecture that opens up the space for communication and social interaction.

These three designs form a model of how we might tackle design in the future and encourage a clear image of 21st-century industry and life.

1. adjust to the specificity of Japanese culture and climate, having learnt from the West
2. develop high-quality function, but minimal product size
3. move on from low-cost production, which Japanese design activity has always pursued in the past
4. create new relationships between humans and products
5. provide technology and engineering with a new goal

It seems as if the key to the future in Japan is hidden behind the design concepts and methodology suggested here. It might even be the key to the future of the world.

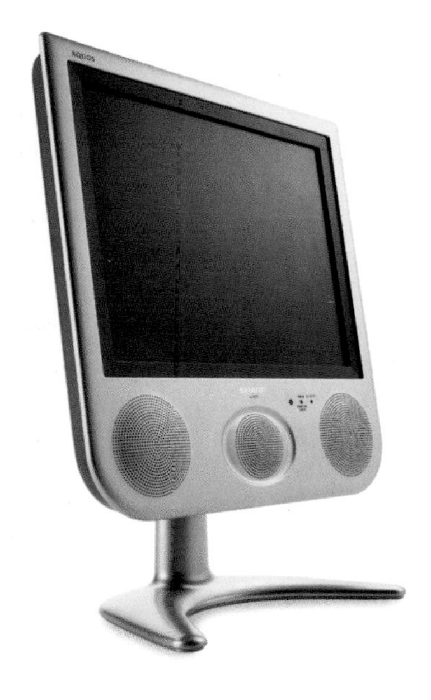

above P-Case CD player/radio from Matsushita (top) and Aquos display monitor from Sharp (bottom)
right Mutech telephone system

past perspectives

HISTORICAL CONTEXT

Like most developed nations, Japan has experienced many politically and culturally tumultuous periods in its long history, but several have particular significance in the development of its social and political structures. In AD 672, a conflict over the succession to the imperial throne led to the Jinshin Disturbance, which for the first time, resulted in the creation of a centralised government. In the 12th century a conflict between the forces of the imperial Taira court in Kyoto and an army from western Japan, split the country once more into a collection of warring regions, controlled by Shoguns and their Samurai warriors. By the 16th century, a state of civil war existed between these powerful, independent warlords, but it was not until loyalist imperialists finally fought against the Shogunate, which ruled the land for two and a half centuries, that the Emperor was reinstated and Japan opened to foreign traders. This decisive conflict was followed by the Meiji restoration in 1868.

Much of what we think of as being distinctively Japanese now is a result of the country's ability to absorb and adapt ideas from other cultures. Korea in the 7th century, China in the 12th century, the Iberians in the 16th century and Western Europe in the 19th century, have all contributed to the rich tapestry of visual, cultural, and technological identity that is modern Japan. But external influences are only part of the story and it is important to recognise the unique character of Japanese creativity. This is becoming more and more evident now, as we understand the social changes that have helped shape Japanese culture.

It is well-known that China and Korea influenced many facets of Japanese life before the Meiji restoration (1868) and in the 19th and 20th centuries after Japan opened itself to the West. Buddhism was introduced to Japan from China and Korea in the 6th century and Shintoism is the peculiarly Japanese adaptation of it. One of its central tenets is that there are spirits present in the natural world, in mountains, rivers, trees and even stones, and it has been argued that Shinto rituals still influence the way society and business is organised in Japan. An appreciation of nature has prevailed in many creative pursuits too, not just in traditional art forms, but also in contemporary design. The legacy of Shintoism may also inform the particularly Japanese skill of imbuing modern technology with an anthropomorphic character.

Confucianism, brought to Japan by the Chinese, was adapted by the Samurai to support notions of brotherhood, morality and honour.

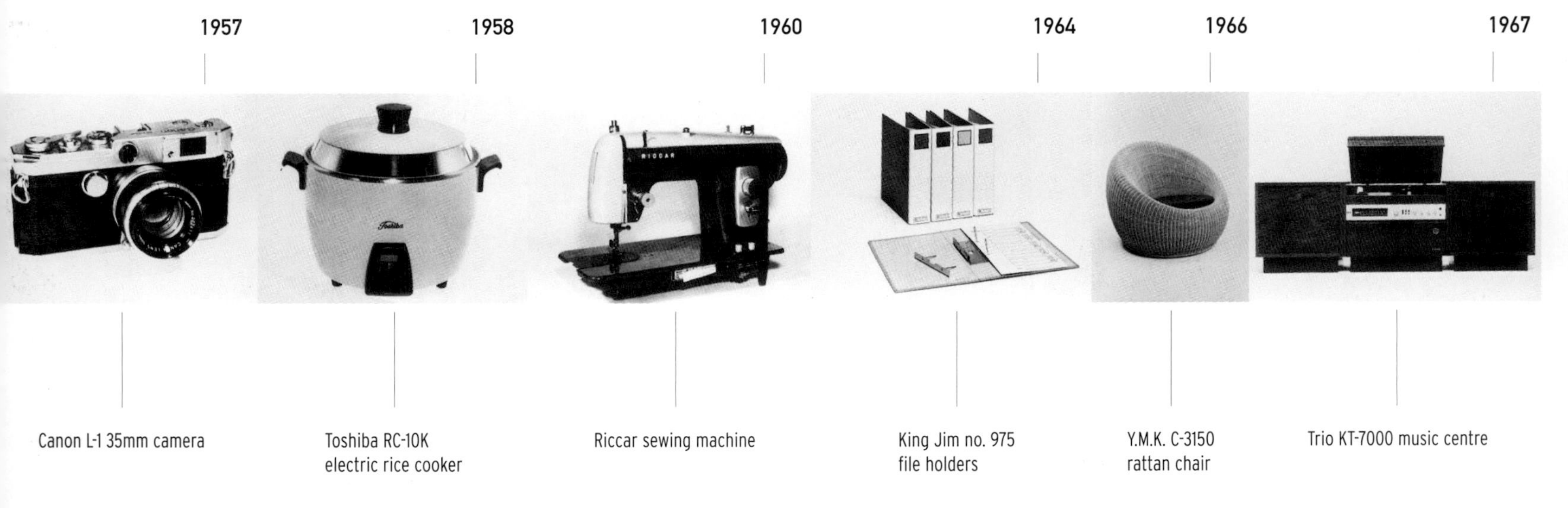

Canon L-1 35mm camera

Toshiba RC-10K electric rice cooker

Riccar sewing machine

King Jim no. 975 file holders

Y.M.K. C-3150 rattan chair

Trio KT-7000 music centre

In some ways, modern Japanese society retains threads of these codes of behaviour in its adherence to ideas relating to hierarchy, bureaucracy, centralised government and group-oriented behaviour. During the 13th century, a further important ingredient in the modern Japanese psyche first appeared. Introduced from China, Zen prizes meditation and physical training to help its followers cope with the rigours of daily life. While a strict discipline was central to the Samurai psychology, the spirit of Zen manifested itself in beautiful buildings and gardens, built for meditation. Meanwhile, social responsibility, still a feature of contemporary Japanese society, has its origins in Samurai culture too. Shoguns imposed strict punishments on lawbreakers and their families for any misdemeanours, which bred a strong sense of collective responsibility that still exists today.

It was during the 18th and 19th centuries that the Japanese first became aware of Western machines and devices, such as steam engines, ships, industrial machinery and domestic appliances (such as they were), and this era is often referred to as the period of 'civilisation and enlightenment'. One clan in Japan even bought firearms from the British to use against the Shoguns during the battles before the Meiji restoration.

It is often argued that the period after Japan opened itself to the West was the beginning of the country's modern industrial age. The wave of technology that arrived from Europe came at a time of economic and political upheaval, when central government was still ill-equipped to impose conditions for change on a national scale. In many ways, it was the Japanese people themselves who seized the initiative. By the end of the 19th century, Japan was training the largest number of scientists and technicians anywhere in the world, and the concept of invention and interpretation gripped this machine-loving people.

This was a period of great global change and invention and Japan's role in it is often underestimated. For instance, western history books routinely state that Otto Lilienthal from Germany made the first unpowered flight in 1891, while the Wright brothers from the US recorded the first powered flight in 1903. But there are also claims that it was Sachikichi Bizenya from Okayama who, in 1804, became the first man to achieve flight with his own glider. (Unfortunately it is thought that Bizenya may have been beheaded by a Shogun, which of course meant an abrupt end to his research.) And it was Cyuhachi Ninomiya, also from Okayama, who in 1893 succeeded in

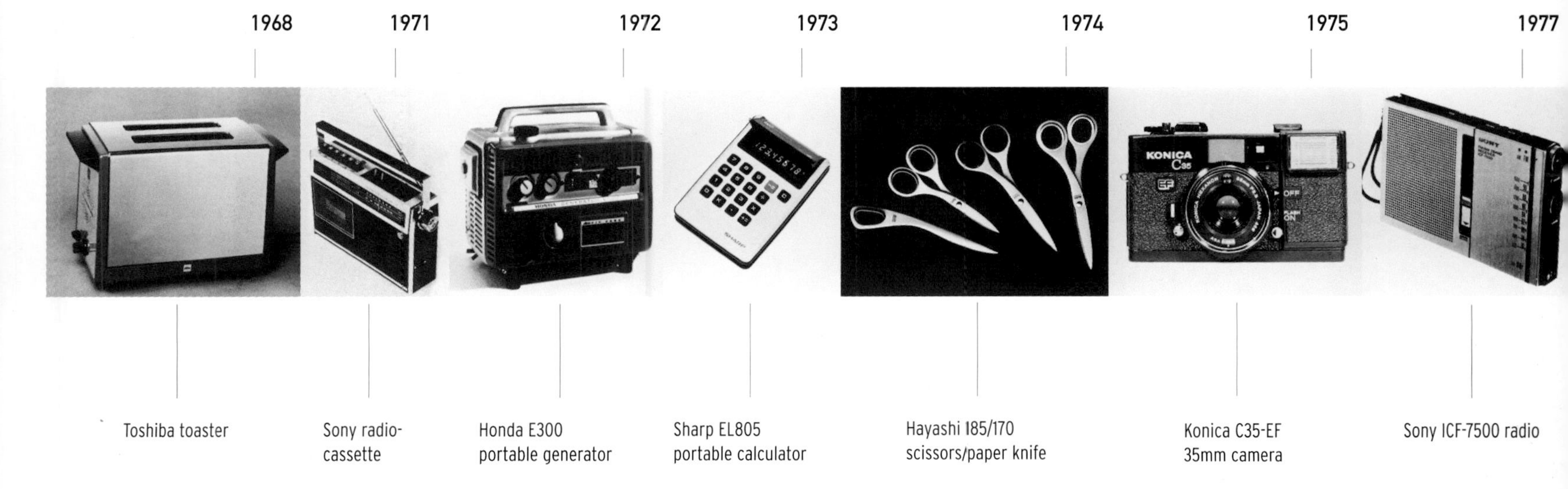

Toshiba toaster

Sony radio-cassette

Honda E300 portable generator

Sharp EL805 portable calculator

Hayashi 185/170 scissors/paper knife

Konica C35-EF 35mm camera

Sony ICF-7500 radio

making the world's first powered flight with a (model) aeroplane. Ninomiya was prevented from developing his ideas by the Japanese military, and when he read about the Wright brothers' subsequent successes with a manned powered flight, he was sufficiently discouraged to give up his quest.

The ready embrace of technology and of contemporary Japanese pre-eminence in manufacturing can be traced through some specific examples from the 1930s. At that time, one of the most popular hobbies for children involved building radios, a kind of home industry that resulted in many related scientific and technological breakthroughs during the period. For instance, the world's first pocket-sized transistor radio, the TR55 of 1955, was developed by Masaru Ibuka and Akio Morita, two visionary engineers working in a humble environment. The pair went on to found Sony Corporation. The seeds of their dominance of world electronics markets were sown in this post-war period, with government incentives encouraging the development of science and technology-based enterprise.

The development of the semiconductor, derived from the integrated circuit, which is in turn derived from the transistor, comes from equally modest beginnings. It is a powerful example of the linear development of one technology leading to dominance in another. Semiconductors have made a profound impression on the modern world, from cars and cameras to phones and computers, and it is certainly not by accident or good fortune alone that Japan leads the world in these industries today.

After the Second World War, US and European businesses and manufacturing models were adopted, adapted and re-exported globally. However, through exploring examples of Japanese artifacts and products from the more distant and recent past – from 19th century Samurai swords to 21st century Global knives; from the Shinkansen to Sony; from Manga cartoon characters to Aibo and ASIMO robots – it is possible to chart the emergence and increasing influence of Japanese manufacturing and culture through its product developments and attitudes to technology.

Following the Second World War, the growth and development of Japan through the changes that took place in its social and business organisation imposed in the name of democracy, at first provided a comfortable sense of collective endeavour. However, a generation later, business success and increased wealth and freedom bred a stronger instinct for personal identity. This fundamental change has not been without some sacrifices. There is a feeling, particularly among older Japanese, that the valuable attributes of conformity, selflessness and social responsibility are being lost, replaced by a more selfish desire for individuality, personal recognition and personal achievement. There is a feeling that certain fundamental qualities of being Japanese – its traditions, sensibilities,

1979 | 1981 | 1982 | 1984 | 1985 | 1989

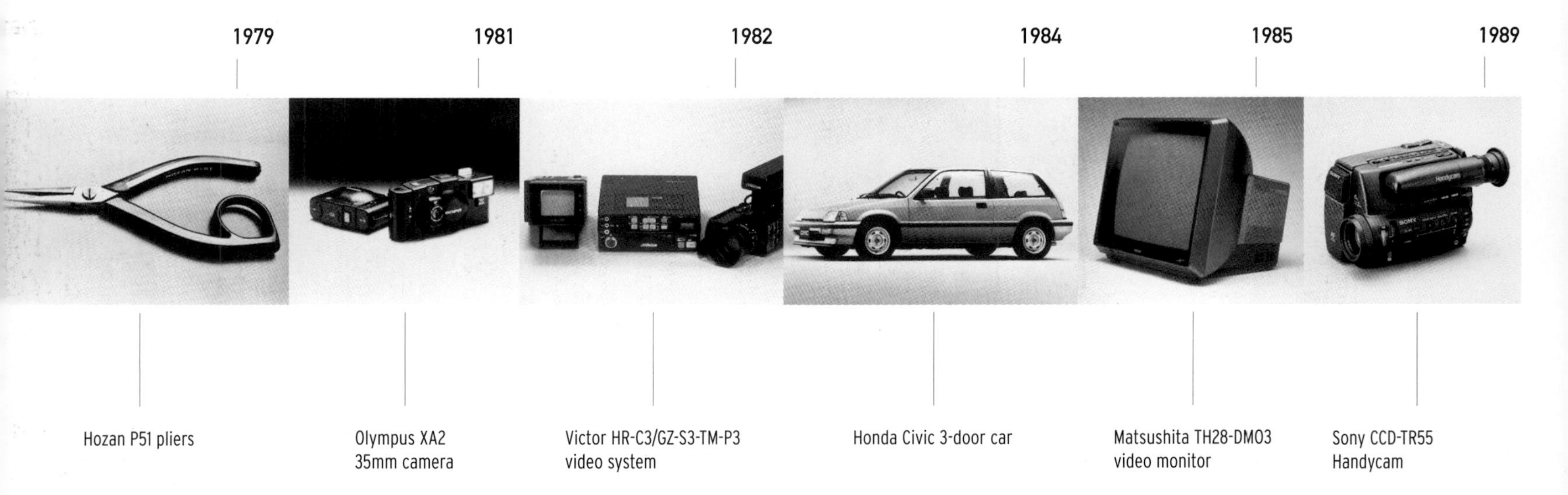

Hozan P51 pliers

Olympus XA2
35mm camera

Victor HR-C3/GZ-S3-TM-P3
video system

Honda Civic 3-door car

Matsushita TH28-DM03
video monitor

Sony CCD-TR55
Handycam

morals, manners and an independent spirit of nationhood that matured through a long history of self-reliance – are disappearing in the space of just a few decades.

It is generally appreciated that throughout history, nations and peoples are pulled between opposing extremes: between phases of imperialist expansion or cultural stagnation, federalism or collectivism, the free market or central control. There is often a choice between the secure and warm embrace of tradition and conservatism or the cold and less certain hand of progress and technology. Whichever direction prevails, its influence will be strong if and when a nation takes part in the 'relay-race' of global economic and cultural importance. During the last 150 years, since the Meiji restoration in Japan, the baton of global influence has passed through the hands of Victorian Britain, inter- and post-war America, arguably, to 1970s Germany and then in the late 20th century to Japan. However, in what is often referred to in Japan today as the post-economic bubble, the realisation that the next recipient, new-millennial China, is already tugging the baton from Japan's grasp is forcing the Japanese to rethink and readapt once again to the realities of a new age.

Eras in Japanese History
(up to the Meiji period, eras were usually named after the capital cities of the time)

10,000 BC — Jomon
200 BC — Yayoi
250 — Kofun
552 — Asuka
710 — Nara
794 — Asuka
1185 — Kamakura
(1192 saw the beginning of the rule of the Shogun or daimyo, meaning 'great names')
1336 — Muromachi
1573 — Azuchi-Momoyama
1615 — Edo
(1603-1868 Japan was ruled by the Tokugawa shoguns)
1868 — Meiji
(Reinstatement of the imperial system, the Meiji restoration)
1912 — Taisho
1926 — Showa
1989 — Heisei

1992 — Komatsu PC50UU-2A excavator

1993 — East Japan Railway Co. Series 209 commuter train

1994 — Grand Bleu Inc. Fieno aqualung

1998 — Bridgestone Transit T20SCX bicycle

1999 — Sony Aibo ERS-110 entertainment robot

2000 — Miyake Design Studio A-Poc design method

2001 — Toyo Ito and Associates Sendai Mediatheque

credits

INTRODUCTION
Photos: p.2 © Canon Inc., p.3 © NEC Corporation, p.4 © Sony Corporation, all photos pp.6-9 © Ian McKinnell, p.10 Game Boy © Nintendo, Tamagotchi © Ian McKinnell, PlayStation © Sony Corporation, p.11 © NEC Corporation, p.13 © Ian McKinnell

CHAPTER 1: ARTIFICIAL EMOTION
Photos pp.14-19 © Ian McKinnell
ASIMO robot pp. 20-25
Company: Honda R&D Co., Ltd Design: Akio Koike, Takeshi Koshiishi, Jun Ito Photos: © Honda R&D Co., Ltd
www.honda.co.jp www.world.honda.com/ASIMO

Aibo robots pp. 26-35
Company: Sony Corporation Design: Hajime Sorayama, Yoshiaki Kumagai, Tsutomu Tsuchiya, Toshio Iribe Photos: © Sony Corporation, except p.29 © Kozo Takayama
www.sony.co.jp www.aibo.com

SDR-3X/4X robots pp. 36-37
Company: Sony Corporation Design: Tatsuzo Ishida, Yoshihiro Kuroki, Sony Corporation Photos: © Sony Corporation
www.sony.co.jp

Q-taro robot pp. 38-39
Company: Sony Corporation Design: Sony Corporation
Photo: © Sony Corporation www.sony.co.jp

PaPeRo robot pp. 40-43
Company: NEC Corporation Design: Yoshihito Fujita, Keigo Kawasaki, Akira Sakai, Kenichi Yoshikawa, Junichi Osada, Aya Yamazaki, NEC Design Ltd. Photos: © NEC Corporation
www.nec.co.jp www.incx.nec.co.jp/robot/PapeRo/english/p-index.html

CHAPTER 2: PERFECTING PERFECTION
Photo p.44 © Ian McKinnell
SFF-2 bow pp. 46-47
Company: Yamaha Corporation Design: Yamaha Corporation
Photo: © Yamaha Corporation
www.yamaha.com www.yamaha.co.jp/english/products/archery

SX-4 Supercomputer pp. 48-49
Company: NEC Corporation Design: Keigo Kawasaki, NEC Design Ltd.
Photo: © NEC Corporation
www.nec.co.jp www.sw.nec.co.jp/hpc/sx-e/Products/sx-4.html

Stella FW2500S fishing reel pp.50-53
Company: Shimano Inc. Design: Shimano Inc.
Photos: © Shimano Inc.
www.shimano.co.jp http://fish.shimano.com/spinning/stella/stella.html

7700 Carbon wheels pp. 54-55
Company: Shimano Inc. Design: Shimano Inc.
Photos: © Shimano Inc.
www.shimano.co.jp http://bike.shimano.com/wheels/road/index.asp

645NII camera pp. 56-61
Company: Asahi Optical Co., Ltd. Design: Gen.Ichiro Ishii
Photos: © Asahi Optical Co., Ltd
www.pentax.co.jp www.pentax.com/products/cameras

Rivaldo Wave Cup football boots pp. 62-65
Company: Mizuno Corporation Design: Takeshi Oorei
Photos: © Mizuno Corporation www.mizunosoccer.com
www.mizunoeurope.com/index_soccer_footwear.html

Potenza Formula One tyres pp. 66-69
Company: Bridgestone Corporation Design: Bridgestone Corporation
Photos: © Bridgestone Corporation
www.bridgestone.co.jp www.bridgestone.co.jp/en/2002/index.html

Shinkansen pp. 70-73
Company: JR Central/West Design: A&F Corporation
Photo: © A&F Corporation www.aandf.co.jp
www.railway-technology.com/projects/shinkansen

CHAPTER 3: THE BEAUTY AND THE BEST
Photo p.74 © Ian McKinnell
EOS-1D digital camera pp. 76-79
Company: Canon Inc. Design: Yasuhiro Morishita
Photos: © Canon Inc.
www.canon.co.jp www.canon.co.uk/eoscameras

Silent Instruments pp. 80-83
Company: Yamaha Corporation Design: Yamaha Corporation
Photos: © Yamaha Corporation
www.yamaha.co.jp www.yamaha.com/cgi-win/webcgi.exe/gss100009

RCV211V and Xasis motorcycles pp. 84-89
Company: Honda Motor Co., Ltd Design: Honda Motor Co., Ltd
Photos: © Honda Motor Co., Ltd
www.honda.co.jp www.honda-racing.co.uk/motogp/machine.php

Aquos display monitors/televisions pp. 90-93
Company: Sharp Corporation Design: Toshiyuki Kita, Sharp Corporation Photos: p.90 © Luigi Sciuccati, pp.91-93 Sharp Corporation www.sharp.co.jp
www.sharp-usa.com/sharp-usa/showcase

CHAPTER 4: MINIATURISM
Photo p.94 © Canon Inc.
IXY APS and digital cameras pp. 96-101
Company: Canon Inc. Design: Yasushi Shiotani (IXY APS) Seiichi Omino (IXY digital)
Photos: © Canon Inc.
www.canon.co.jp www.canon.co.uk/apscameras
www.canon.co.uk/digitalcameras

Mebius Muramasa laptop PC pp. 102-105
Company: Sharp Corporation Design: Mitsuo Kawa, Sharp Corporation
Photos: © Sharp Corporation,
www.sharp.co.jp www.sharp.co.uk/muramasa

EMRoS robot pp. 106-109
Company: Seiko Epson Corporation. Design: Noritaka Uchibori
Photos: © Seiko Epson Corporation, except p.107 © Tsutomo Sakamoto
www.epson.co.jp www.epson.co.jp/osirase/1996/1118.htm

F601 digital camera pp. 110-111
Company: Fuji Photo Film Co., Ltd Design: Design Centre, Fuji Photo Film Co., Ltd. Photos: © Fuji Photo Film Co., Ltd
www.fujifilm.co.jp www.fujifilm.com/JSP/fuji/epartners

G-Shock watches pp. 112-115
Company: Casio Computer Co., Ltd Design: Casio Computer Co., Ltd
Photos: © Casio Computer Co., Ltd www.casio.co.jp
www.casio.com/watches

N502i i-mode phones pp. 116-119
Company: NEC Corporation Design: Chiaki Suzuki
Photos: © NEC Design Ltd, Na-cord p.118 Design: Maywa Denki, photo © Ian McKinnell www.nec.co.jp

CHAPTER 5: CONSUMER WORLD CHAMPIONS
Photos p.120 © Sony Corporation
PlayStation and PlayStation 2 pp. 122-125
Company: Sony Corporation Design: Teiyu Goto
Photos: © Sony Corporation. www.sony.playstation.com

TR55 and IP7 Handycams pp. 126-127
Company: Sony Corporation Design: Sony Corporation
Photos: © Sony Corporation www.sony.com

Cheki instant camera pp. 128-129
Company: Fuji Photo Film Co., Ltd Design: Design Centre, Fuji Photo Film Co., Ltd Photo: © Fuji Photo Film Co., Ltd www.fujifilm.co.jp

Game Boy, Game Boy Advance, N64, GameCube pp. 130-135
Company: Nintendo Design: Nintendo
Photos: © Nintendo
www.nintendo.co.jp www.nintendo.com/systems/index.jsp

Soft Iron, Electric Bucket pp. 136-137
Company: Matsushita Electrical Industrial Co., Ltd
Design: Fujita Kazuhiro, Ueda Yoshiaki, Hirotoshi Hada, Kazuya Takahashi Photos: © Matsushita Electrical Industrial Co., Ltd
www.national.co.jp www.design.panasonic.co.jp

WiLL pp. 138-141
Companies: Toyota, Matsushita, Kao, Kinki Nihon, Asahi, Kokuyo, Design: various Photos: © WiLL www.willshop.co.jp

CHAPTER 6: TRANSFORMATIONS
Photo p.142 © Honda Motor Co., Ltd
WX7/5 MIDI Instruments, Silent Taishogoto pp. 144-147
Company: Yamaha Corporation Design: Yasuhiro Kira, Yamaha Corporation (WX7/5), Yamaha Corporation (Silent Taishogoto)
Photos: © Yamaha Corporation
www.yamaha.com www.yamaha.co.jp

Global kitchen knives pp. 148-153
Company: Yoshikin Metal Co., Ltd Design: Komin Yamada
Photos: © Ian McKinnell Drawing p.150: Komin Yamada
www.yoshikin.co.jp

Film with lens pp. 154-159
Company: Fuji Photo Film Co., Ltd Design: Design Centre, Fuji Photo Film Co., Ltd Photos: © Ian McKinnell except p.155 © Fuji Photo Film Co., Ltd www.fujifilm.co.jp

Nexave deraillier pp. 160-163
Company: Shimano Inc. Design: Shimano Inc.
Photos: © Shimano Inc. www.shimano.co.jp
http://bike.shimano.com/comfort/nexavet400/index.asp

Caixa Unibox concept car pp. 164-165
Company: Honda Motor Co., Ltd Design: Honda Motor Co., Ltd
Photos: © Honda Motor Co., Ltd www.honda.co.jp

NEW JAPANESE DESIGN pp. 166-169
Photos: p.166 Sendai Mediatheque © Toyo Ito and Associates, A-Poc © Nacàsa & Partners, p.168 Aibo robot © Sony Corporation, p.169 Aquos monitor © Luigi Sciuccati

PAST PERSPECTIVES: HISTORICAL CONTEXT pp. 170-173
All images courtesy of JIDPO (Japan Industrial Design Promotion Organisation)

Photo p.176 © Sony Corporation
Composite images pp. 34-35, 97, 103, 120, 133, 137, 156-157 by TKO
Drawings pp. 23, 72-73 by TKO

Back cover also features Nikon F series.

Other useful websites:
www.tkodesign.co.uk
www.bandai.com/products/gundam
www.takaratoys.co.jp/aquaroid
www.maywadenki.com
www.jidpo.or.jp
www.jida.or.jp
www.axis.co.jp
www.androidworld.com
www.robofesta.net
www.sonet.co.jp

acknowledgements

I am immensely grateful to all the companies who have contributed to this book, without their help and cooperation this book would literally not have been possible.

My heartfelt thanks and very best wishes to the following individuals:

A&F Corporation Tetsuo Fukuda
Bridgestone Sarah French
Canon Mami Urimoto, Masahiro Kando, Atsushi Ueno, Richard Berger, Seiichi Omino, Yasushi Shiotani, Yuzo Oka, Yasuhiro Morishita
Casio (Japan) Toshiya Ando, Toshihiro Watanabe
Casio (Europe) Michael Williamson, Gordon Dickens, Lisa McGonigle
Epson Noritaka Uchibori
Fujifilm Hiroshi Fukuda, Chosei Sawa, Makoto Isozaki, Koji Kamitani
Fujitsu Mark Friesen, Yuko Shimura
Hitachi Eiichi Kubota
Honda (ASIMO) Yuji Hatano, Akio Kioke, Takesh Koshiishi
Honda R&D Shunsuke Yoshida, Misa Imazawa, Jun Tomizawa
JIDPO Yuichi Yamada, Ai Hasegawa
Kokuyo Shoko Tomonaga, Nobutaka Sakai, Akihiro Taketsuna, Motoo Kumagai
Matsushita Tetsuya Imamura
Mizuno Noboru Kohno, Tadashi Matsuda, Takeshi Oorei,
NEC Akira Sakai, Keigo Kawasaki, Yuko Ogawara, Chiaki Suzuki, Toru Ichihashi, Noriko Kitaoka
NEC Europe Claire McSharry
Nintendo (Cake Media) Jennie Kong
Pentax Asako Hasegawa, Gen.ichiro Ishii, Kazumichi Eguchi
Sharp Mikio Yamashita
Shimano Seiji Myojo, Masao Kojima, Masakazu Iwabuchi, Etsuyoshi Watarai, Kage Nishimura
Sony (Japan) Yoshinao Kambe, Ryosuke 'Ricky' Yokota, Shoji Yoshi, Yuka Takeda Aki Shimazu, Shinji Obana, Mina Naito, Nanako Kato
Sony Corporation Masahiro Nagakubo, Atsushi Suzuki, Yuko Tsukamoto
Sony SCEE Imogen Baker
WiLL Masaki Ono
Yamaha Yasuhiro Kira
Yoshikin Kazutaka Hayano, Hideyo Watanabe, Satoshi Ono, Natsue Shinoda, Kenji Shibuya
British Embassy/Trade Partners Donald Spivey, Alison Scott

Eizi Hayashi
Shinya Iwakura
Youzaburo Yamashita
Yuko Iwakura
Ken Mars Sekiguchi
Itaru Sugino
Ai Sugino
Noriji Sato
Katsutoshi Ishibashi
Eichi Kono
Hilary Knight
Ian McKinnell
Komin Yamada
Masayuki Kurakata
Yasuko Kurakata
Mariko Hamaguchi
Akiko Miyahara
Yuki Hiroma
Rika Yamanaka

Very special thanks to Annie Gardener and Rochelle Kleinberg at TKO, Jo Lightfoot, Simon Cowell and Zoe Antoniou at Laurence King.

overleaf Aibo ERS-312 Latte and ERS-311 Macaron